IUPAB Biophysics Series
sponsored by
The International Union of Pure and Applied Biophysics

Fundamentals of light microscopy

IUPAB Biophysics Series
sponsored by
The International Union of Pure and Applied Biophysics
Editors:
Franklin Hutchinson
Yale University
Watson Fuller
University of Keele
Lorin J. Mullins
University of Maryland

1 Walter Harm: Biological effects of ultraviolet radiation
2 Stanley G. Schultz: Basic principles of membrane transport
3 Edward G. Richards: An introduction to the physical properties of large molecules in solution
4 Roderick K. Clayton: Photosynthesis: physical mechanisms and chemical patterns
5 E. G. Fein and E. Szuts: Photoreceptors: their role in vision
6 Michael Spencer: Fundamentals of light microscopy

THE

MICROSCOPE
Made Easy:

OR,

I. The *Nature*, *Uses*, and *Magnifying Powers* of the best Kinds of MICROSCOPES *Described*, *Calculated*, and *Explained*:

FOR THE

Instruction of such, particularly, as desire to search into the WONDERS of the *Minute Creation*, tho' they are not acquainted with *Optics*.

Together with

Full Directions how to *prepare*, *apply*, *examine*, and *preserve* all Sorts of OBJECTS, and proper Cautions to be observed in viewing them.

II. An Account of what surprizing *Discoveries* have been already made by the MICROSCOPE: With useful Reflections on them.

AND ALSO

A great Variety of new *Experiments* and *Observations*, pointing out many uncommon Subjects for the Examination of the CURIOUS.

By *HENRY BAKER*, Fellow of the *Royal Society*, and Member of the Society of *Antiquaries*, in *London*.

Illustrated with COPPER PLATES.

Rerum Natura nusquam magis quam in Minimis tota est.
PLIN. Hist. Nat. Lib. XI. c. 2.

LONDON:
Printed for R. DODSLEY, at *Tully's Head* in *Pall-Mall*.
M. DCC. XLII.

CHAP. XV.

Cautions in viewing Objects.

BEware of determining and declaring your Opinion suddenly on any Object; for Imagination often gets the Start of Judgment, and makes People believe they see Things, which better Observations will convince them could not possibly be seen: therefore assert nothing till after repeated Experiments and Examinations in all Lights and in all Positions.

When you employ the Microscope, shake off all Prejudice, nor harbour any favourite Opinions; for, if you do, 'tis not unlikely Fancy will betray you into Error, and make you think you see what you would wish to see.

Remember that Truth alone is the Matter you are in search after; and if you have been mistaken, let not Vanity seduce you to persist in your Mistake.

Pass no Judgment upon Things over-extended by Force, or contracted by Dryness, or in any Manner out of their natural State, without making suitable Allowances.

There is no Advantage in examining any Object with a greater Magnifier than what shews the same distinctly; and therefore, if you can see it well with the third or fourth Glass,

Frontispiece. Advice of an eighteenth-century microscopist.

Fundamentals of light microscopy

MICHAEL SPENCER

Department of Biophysics
University of London, King's College

CAMBRIDGE UNIVERSITY PRESS
CAMBRIDGE
LONDON NEW YORK NEW ROCHELLE
MELBOURNE SYDNEY

CAMBRIDGE UNIVERSITY PRESS
Cambridge, New York, Melbourne, Madrid, Cape Town, Singapore, São Paulo, Delhi

Cambridge University Press
The Edinburgh Building, Cambridge CB2 8RU, UK

Published in the United States of America by Cambridge University Press, New York

www.cambridge.org
Information on this title: www.cambridge.org/9780521289672

© Cambridge University Press 1982

This publication is in copyright. Subject to statutory exception
and to the provisions of relevant collective licensing agreements,
no reproduction of any part may take place without the written
permission of Cambridge University Press.

First published 1982
Re-issued in this digitally printed version 2009

A catalogue record for this publication is available from the British Library

Library of Congress Catalogue Card Number: 82–1163

ISBN 978-0-521-24794-8 hardback
ISBN 978-0-521-28967-2 paperback

CONTENTS

PREFACE

Despite enormous advances in electron microscopy (most recently in the scanning of surface structure) the light microscope is far from obsolete, and new applications are still being invented. The most striking of these have been in the biomedical area, where fluorescence microscopy in particular has opened up whole new fields of research. Phase-contrast and the newer interference-contrast microscopy remain unrivalled for observing living cells, while quantitative interference techniques and polarizing microscopy find application in the industrial field as well. The light microscope is probably the biophysical tool most widely used by non-physicists.

The electron microscope is essential for visualizing detail at the molecular level, but it cannot match the ability of the light microscope to focus at different levels within a three-dimensional object; nor can it offer the range of specific staining techniques available to the light microscopist. In biology the data from all kinds of microscopy are needed to study different levels of organization. There are, however, inherent problems of limited resolution in light microscopy, and a possibility of optical artifacts, that are just as real today as they were two centuries ago (Frontispiece). The aim of this book is to explain to the practical microscopist in mainly non-mathematical terms the basic principles underlying each branch of light microscopy, so that he or she can get the best out of an instrument and avoid misleading results. Some mathematical treatments have been collected in the Appendix, but these should be regarded as strictly optional.

The book should also help the research worker in choosing what is often a fairly expensive piece of apparatus. It is designed particularly for biomedical students, and for all those preparing for professional examinations in microscopy. It arose out of a course of lectures given for some years past in King's College. Our students have mainly been undergraduates, but the course has always been open to postgraduate students and research workers from other institutions. We have therefore had the benefit of interacting with enthusiasts from a wide range of backgrounds, some of whom had never studied physics before and needed a concise but intellectually-satisfying explanation of the basic principles. At the same time, many undergraduates who had studied physics at 'A' level were glad to be reminded of what they had once learnt but since forgotten.

A few topics have deliberately been excluded because I believe them to be better left to the specialized textbooks that are available; they include staining techniques, micrometry, stereology, and the use of various scanning devices for making quantitative measurements. Scanners rely on the integration of the light transmitted by a specimen when traversed by a 'flying spot'. Apart from this they involve no new principles, and have the same basic optical system as an ordinary microscope. Their high cost means that only a research worker is likely to have access to them, and the manufacturers of such devices provide adequate instructions. I have, however, included a brief discussion of photomicrography; the arrangements for taking pictures vary between microscopes, but the basic principles are the same as in any application of photography.

For help in providing material and making suggestions I am particularly indebted to the late Howard Davies, who was instrumental in setting up the original microscopy course, and to Tom Cavalier-Smith who currently organizes it. I am also grateful to many colleagues and others who have read and criticized the manuscript. The half-tone illustrations owe much to the skill of Zoltan Gabor, and originals were kindly provided by Donald Olins (Frontispiece), Michael Dickens (Fig. 13), John Couch and Ed O'Brien (Fig. 21), Derek Back and Howard Davies (Figs 23 and 26), Clive Thomas (Fig. 30), and John Couchman (Fig. 43). The Frontispiece was previously reproduced in Ts'o, P.O.P. (ed.) (1977) *The molecular biology of the mammalian genetic apparatus, Vol. I* (North-Holland, Amsterdam).

1 Microscope alignment

We shall see in later chapters why there is a theoretical limit to the ability of a light microscope to see fine detail, or to distinguish closely-spaced objects from each other. Many biological objects are of a size close to this resolution limit, which is defined as the smallest detectable separation of two point objects. It is clearly important that the instrument is set up to approach the limit as closely as possible. With good design this can be achieved much more readily than for an electron microscope, whose resolution is often far short of its theoretical limit. A student who has used only the preset microscopes provided in a practical class may not appreciate what a difference it makes to have the optical system properly aligned. In particular, there are theoretical grounds for the belief that the form of the *illumination* affects the resolution. Here we consider how this and other equally important factors are optimized when setting up a microscope. The further alignment of specialized instruments such as the phase microscope is discussed in later chapters.

Basic geometrical optics

This section is for those who would like to be reminded of the fundamental properties of lenses. Any microscope can be considered as equivalent to only two simple, converging (magnifying) lenses, representing the objective and the eyepiece. The condenser is equivalent to a third lens, and a fourth is used to achieve *Köhler illumination* (see next section). In practice, of course, each component in the system contains many optical elements, but this does not concern us now.

A *converging* lens has two *foci* (F_1 and F_2 in Fig. 1(a)) which are the meeting points of parallel rays of light striking the opposite side of the lens. The *focal length*, f, is the distance from the centre of the lens to either of the two foci. Fig. 1(a) also illustrates why the *depth of focus* (related to the distance over which the rays converge more or less to a point) is greater when the lens aperture is restricted by a *stop*; on either side of F_2 the light forms a blurred circle instead of a point image, but with a stop in position one can move further away from F_2 before the effect is noticeable. This is why there is no need to adjust the focus of a cheap camera whose aperture is permanently restricted.

A converging lens can produce two kinds of image. A *real image* (one that can be formed on a screen or a photographic film) arises when the object is placed so that its distance from the lens is greater than the focal length. There are mathematical formulae (Fig. 1(b)) for predicting both the position and the magnification of the image in terms of u (distance of object from lens), v (distance of image from lens), and f. However, there is also a very simple graphical method of determining them. From a point on the object draw two rays, one of which travels parallel to the lens axis OL, while the other passes through the centre of the lens. The first ray must bend so as to pass through the focus at point F_2 on the other side of the lens, while the second ray is not changed in direction because the centre of the lens behaves like a parallel-sided block of glass. The image is formed, upside down, where these two rays intersect.

For a given depth of focus in the image there will be a proportional *depth of field* in the object, for which a tolerably-sharp image is obtained. Low-power microscope objectives have in general a greater depth of field than high-power ones.

A *virtual image* (Fig. 1(c)) is formed by a converging lens when the object is placed so that its distance from the centre of the lens is *less* than the focal length. Rays drawn as before will no longer converge to a point; however, a detector (such as the eye or a camera) that incorporates an additional converging lens can form a real image. The image *appears* to be situated on the same side of the lens as the object, and in this case it is not inverted. When viewing by eye, the distance from eye to image must not be less than the minimum distance for comfortable viewing (about 250 mm). For completely relaxed eyes, the virtual image can be 'at infinity' as in Fig. 1(d); this arises when the object lies exactly at the focus of the lens. The eye regards it as a distant object (such as the sun) whose size can only be expressed in terms of an angle β (beta) between the rays reaching the eye from the extreme edges of the object.

This leads to a convention which defines the magnification of a virtual image, and which is commonly used for microscope eyepieces. The magnification is taken as the ratio of the angles β and α (alpha) made at the eye by the image, and by the object when viewed directly at a distance a of about 250 mm. Incidentally, some people find it impossible to relax their eyes completely when using a microscope, and they focus it to form a virtual image quite close to the eye. This can lead to problems when setting up for photography, unless one uses an attachment to form a real image on a ground-glass screen.

An ordinary magnifying glass forms a virtual image, but there is a

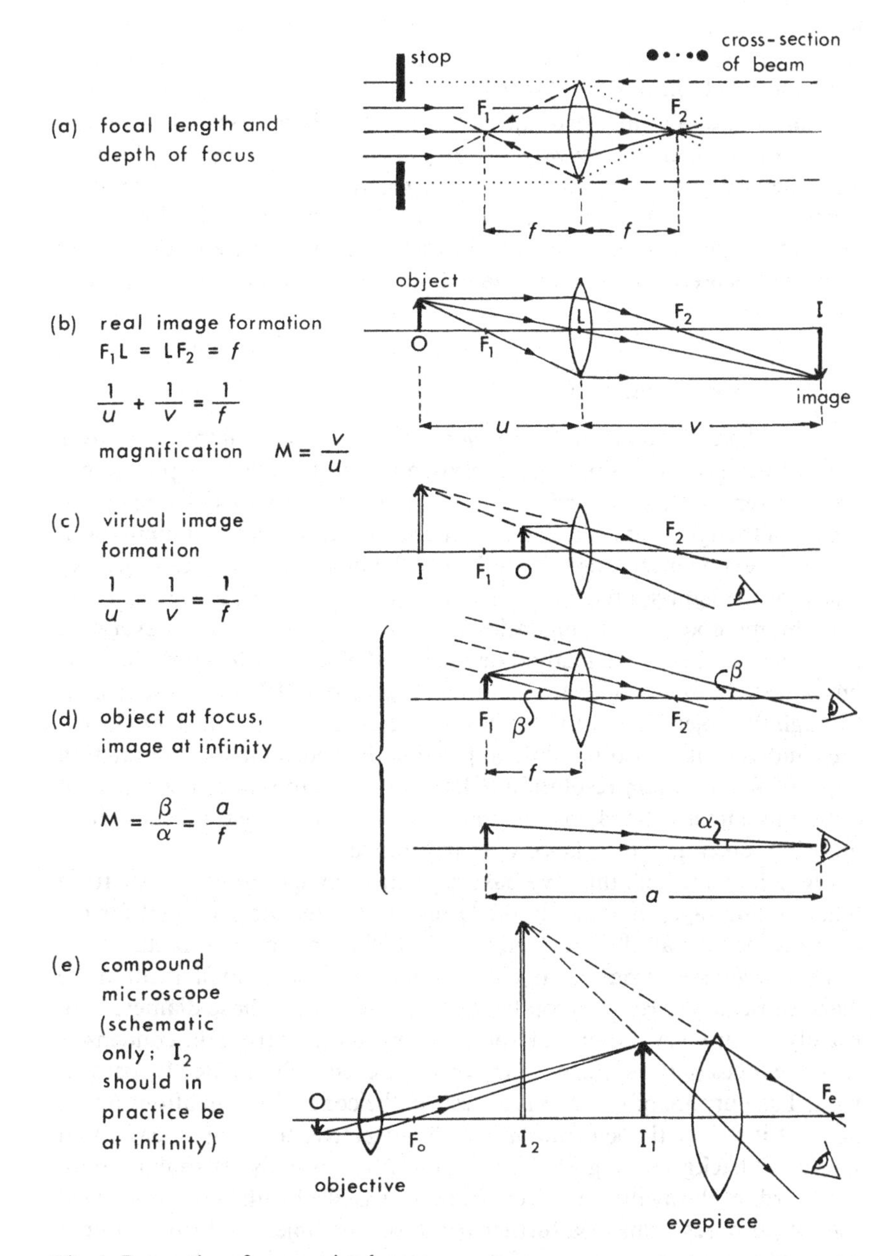

Fig. 1. Properties of converging lenses.

practical limit to the magnification obtainable. Nobody in modern times has matched the achievement of the seventeenth century microscopist van Leeuwenhoek, who obtained the astonishing resolution of about 1.5 micrometres (1 μm $= 10^{-6}$ m) with tiny single lenses, magnifying more than 200 times and specially ground to reduce the aberrations. All modern microscopes are of the *compound* variety (Fig. 1(e)); the objective forms a magnified real image I_1, and the eyepiece generally forms a virtual image I_2 which has been magnified relative to I_1. The total magnification is the product of the magnifications of objective and eyepiece, and the microscope is focused by moving the whole combination relative to the object.

Köhler illumination

There was a violent controversy which lasted until the 1930s over the best means of illuminating a microscope, and the misconceptions that were current then still persist in some modern textbooks. There was a school of thought that insisted on the special virtue of 'critical illumination' in which a condenser lens (placed below the object) is used to project an image of the light source onto the object plane. There is, however, a risk with this method of obtaining rather uneven illumination. Nearly everyone now uses the alternative method devised by Köhler, in which the source is imaged at the focus of the condenser to give parallel (unfocused) light through the object plane. By forming a magnified image of the lamp below the condenser it is also possible to provide the wide cone of illumination required for optimum resolution (Chapter 4), and to reduce the effects of dust and imperfections in the condenser. As a further precaution, a diffusing screen is often placed over the source.

Fig. 2 illustrates all this. We have separated 'image-forming rays' from 'illuminating rays,' but clearly the latter must enter the eye pupil for the image to be seen at all. Let us consider the illuminating rays first.

The *Köhler lens* produces an enlarged image of the lamp filament in the plane of the *aperture stop* (condenser iris). This stop, whose diameter can usually be varied by moving a knob or lever projecting from the condenser, is at one focus of the condenser; the object is evenly illuminated by parallel light. The purpose of the stop is to restrict the cone of illumination so that none of it strikes the lens mounts of the objective; this would lead to an unwanted background 'glare'. The diameter is normally set so that about two-thirds of the available objective aperture is filled with light, though for special purposes it may be further reduced. The objective forms another image of the source (and one of the aperture stop) in its *back focal plane*, and the eyepiece causes the rays to converge again towards the pupil of the eye. In order to test whether the lamp is properly aligned one can use a

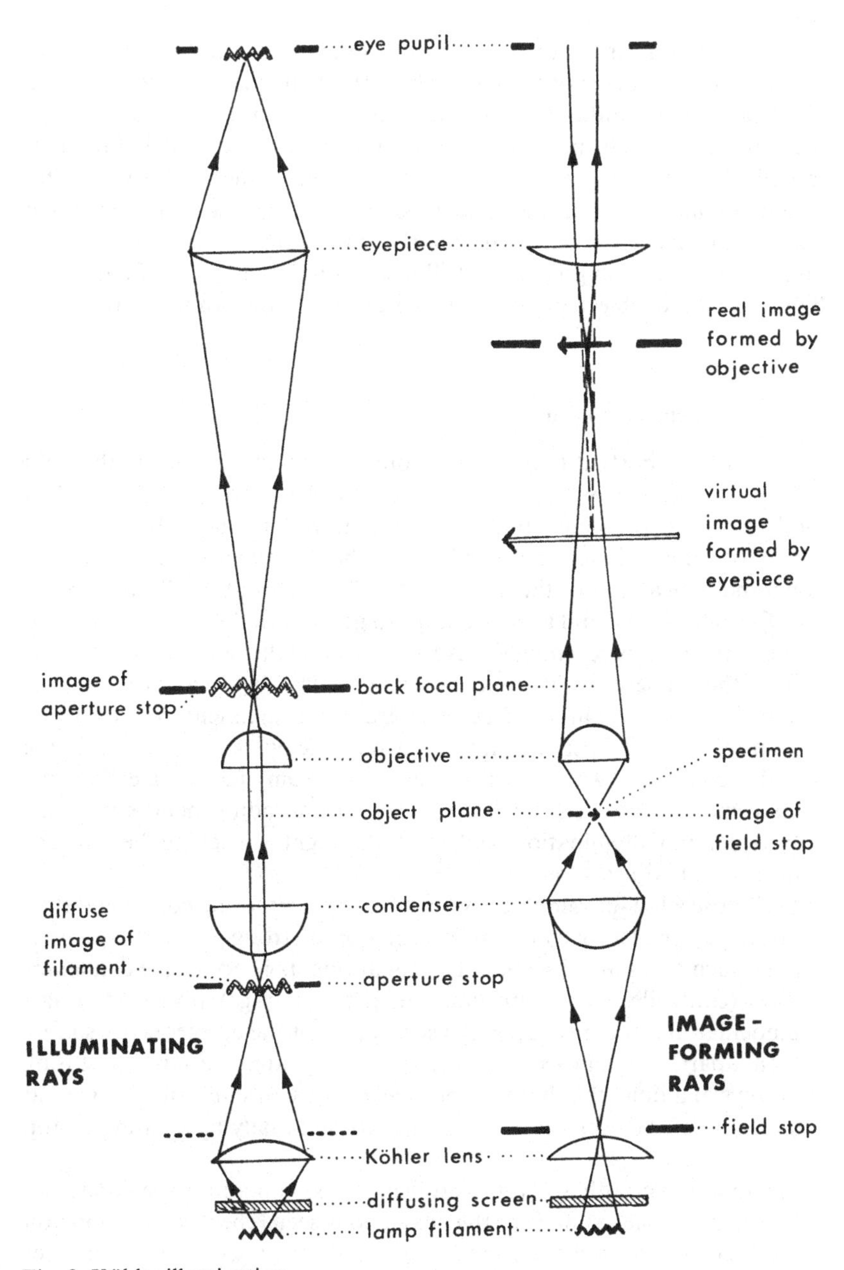

Fig. 2. Köhler illumination.

phase telescope (or a swing-out *Bertrand lens* if fitted) to view the back focal plane.

The image-forming rays are as in Fig. 1(e), with the addition of a set of rays forming an image of the *field stop* in the same plane as the specimen. The field stop is another variable diaphragm, placed some distance below the condenser (on the microscope base in many instruments) and it serves to limit the area illuminated to that actually seen in the field of view. This eliminates unwanted background arising from light scattered by other regions of the specimen – particularly important when viewing faint objects, or when using dark-field illumination (discussed in Chapter 5). Note that each object point receives light from the whole of the lamp filament.

Alignment procedure

It is important to adjust the controls of a microscope in the right sequence, otherwise one setting may spoil another made previously. Most modern microscopes have built-in illumination, but where this is not the case extra precautions are needed to get the illuminating rays travelling down the optical axis of the microscope. The stages best followed are:

1) Open both field and aperture stops to get some light into the system. If very little light is reaching the eyepiece, follow the illumination coming from the condenser with a piece of paper placed between the condenser and the objective, and correct any gross misalignment. With built-in illumination, this is done simply by turning the two screws that move the entire condenser with its associated stops from side to side; in some instruments, however, these are preset and no adjustment is possible. With external illumination, make sure that light is reaching the substage mirror and adjust its setting.

2) Choose a low-power objective (e.g. × 16); next, using coarse and then fine adjustments, focus the microscope on a strongly-absorbing specimen, such as a stained section. Do not at this stage adjust the condenser focus (controlled by a knob under the specimen stage). If you are using a binocular microscope, adjust the separation of the eyepieces (by sliding them apart) to avoid seeing a double image, and make sure (by closing one eye at a time) that both images are in focus; in some microscopes, at least one of the eyepieces can be focused separately by rotating its top lens.

3) Close down both aperture and field stops, and adjust the condenser focus until a sharp image of the field stop is seen (check which stop you are focused on by varying their diameters). Next centre the condenser (where possible) so that the field-stop image is in the middle of the field of view, and open the field stop again to just fill the field.

4) Where a bright-field microscope is being used, close down the aperture stop attached to the condenser. Insert a phase telescope in place of the eyepiece, or a swing-out lens where fitted. If the telescope is adjustable, focus it so that a sharp image of the aperture stop is seen. Now move the controls (where fitted) that centre the aperture stop in relation to the condenser, so as to centre the stop image in the field. Open the stop to fill about two-thirds of the maximum available aperture (determined by opening the stop to show the limit set by the objective). Replace the eyepiece where necessary.

5) Change to a higher power if needed, and repeat adjustments 3 and 4. In the more expensive instruments, the microscope focus will need little adjustment between objectives (which are then described as *parfocal*), but the centring controls will almost certainly need changing as the highest power is approached. The aperture stop should be opened up, but the field stop will need closing because the field of view is smaller for the higher powers. Do not expect the image of the field stop to be as well defined at high power, as most condensers suffer from optical defects (aberrations) that show up under such conditions.

A few elementary precautions are worth noting, especially where photography is contemplated. *Dust* is the great enemy of the microscopist and is very difficult to eliminate entirely, particularly if it has got into the interior of the instrument. Every microscope should have a dust cover over it when not in use. A pressurized can of inert gas (as used by photographers) is useful in removing dust from lens surfaces. Fingermarks will also introduce unwanted background and even affect the resolution. To remove them it is often sufficient to breathe on a lens to deposit 'distilled' water on it, and to wipe the surface with a clean, soft tissue. Before starting work, inspect all surfaces of condenser, objective and eyepiece by reflected light, and blow off any loose particles. Immersion oil can be removed with a small amount of organic solvent, but the manufacturer's handbook should first be consulted; even alcohol may attack the cement holding the objective front lens in place.

Although many modern objectives are spring-loaded, it is still not a good idea to focus down towards the specimen; always set to the *minimum* separation by external inspection and then focus upwards (increasing separation). Contact between objective and slide may well scratch the front lens of the objective and spoil its anti-reflection coating. The same dangers exist for any component or accessory—always put it in a safe place rather than simply leaving it on the bench.

Questions to test your understanding (*answers follow Appendix*)

1. If given only a magnifying glass, a piece of card and a ruler, how could you estimate the focal length of the lens?
2. With the equipment listed in question 1, how could you define the magnification of the lens?
3. When a magnifying glass is used in conjunction with the relaxed eye and an object is placed at the focus of the lens, where is the image formed?
4. If a transparency is illuminated from behind and an image of it projected onto a screen by a lens, what will happen if the illuminating lamp is moved to one side – will the image move, or not?
5. If light is scattered from regions of an object outside the field of view, will spurious images ever be formed? If no spurious images are formed, is the scattered light in any way detrimental?

2 Properties of light

Basic principles – why bother to learn them?

We first have to consider the wave properties of light. Wave theory is essential in explaining how transparent objects are made visible by phase or interference techniques; why there is a limit to the detail that can be seen; how an instrument's performance is tested; and how quantitative measurements are made with polarizing and interference microscopes. Only qualitative staining techniques, where light is simply absorbed by the object, require no further understanding of the process involved – and even then it is a good idea to know enough about optics to optimize the illumination. Staining is not, however, usable when we need to visualize transparent objects without subjecting them to chemical insults, nor can it in general be used for determining molecular orientation or the dry mass of an object.

There is a further bonus for those who master the basic principles of physical optics: the theory applies directly to both X-ray diffraction and optical diffraction, now a popular tool for analyzing electron micrographs. Although this book deals only briefly with such techniques, it should give the reader enough insight to understand how the specialists obtain their information, and how reliable it is.

Waves and particles

It is a long-standing paradox of physics that light waves must sometimes be treated as if they consisted of particles in order to explain their behaviour, while electrons have sometimes to be regarded as waves. However, waves and moving particles have in common the property of carrying *energy* from the source to the observer. The eye is sensitive only to changes in brightness and in colour, so in microscope images the brightness or *intensity* is made to vary across the field of view. As we shall see, many biological objects do not in themselves generate intensity changes, and a special attachment (as in the phase microscope) must be used to do this by making visible the variations in refractive index. The same problem arises in photomicrography.

Properties of waves

A wave is the result of propagating a disturbance from a source, as in the ripples made by a stone dropped in a pond. Energy from the source (for instance, thermal energy in the atoms of a lamp filament) is fed into the waves. In the case of monochromatic waves like those generated by a discharge lamp or a laser, the wave responsible for a given colour (e.g. the intense green line of a mercury lamp) is characterized by a *frequency*, measured in Hertz (cycles per second), which is invariant; an observer anywhere in the path of the wave will measure the same frequency, and the frequency determines the perceived colour of the light. The rule only breaks down where there is rapid relative motion of source and observer, a situation that does not concern us here.

In the generation of a sound wave, some kind of surface vibrates, causing variations in air pressure which travel out from the source. For light waves from a hot filament or a discharge lamp, each atom acts as an independent source – only in a laser do they act in unison. In addition, the energy given out by each atom comes in bursts (quanta), and the timing of the start of each burst is independent of that of the last one. We shall see the relevance of this later.

A light wave, like any electromagnetic wave, is propagated as a fluctuation in electric and magnetic *fields*. An example of a static electric field is shown in Fig. 3, where parallel plates are connected to a battery. Note that the field, unlike the pressure variation in a sound wave, is directional and has a sign as well as a magnitude which can be measured with a suitable instrument; reversing the battery terminals will reverse the direction of the field. In a propagated wave (also shown in Fig. 3) there is a periodic oscillation in electric field along the direction of propagation, and

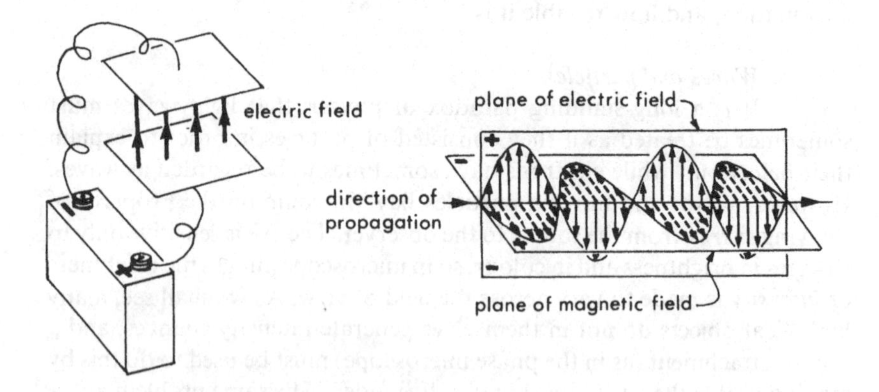

Fig. 3. An electric field and an electromagnetic wave.

an associated magnetic field varying in the same way; since these two fields always go together, we need consider only one of them to define the wave. Note that both fields are directed at right angles to the direction of propagation: the waves are *transverse waves* like the moving ripples on a pond. The wave illustrated, which is *plane polarized*, is characterized by a fixed plane in which the electric field vibrates. There are other types in which the plane is distorted like a twisted ribbon, but they do not concern us here. We shall discuss polarization in more detail in Chapter 7.

The wave shown in Fig. 3 is 'frozen' at an instant in time; in practice the whole system is moving forward at the velocity of light, and an observer who is stationary at a given point will observe variations in field *with time* like the variations with distance in Fig. 3. As with sound waves, the velocity depends on the medium; but unlike sound waves, light can travel through a vacuum and is normally *slowed down* by the presence of matter. The velocity *in vacuo* ('free space') has a value $v_0 = 3 \times 10^8$ m s^{-1} that is independent of all other properties of the wave; light waves and radio waves travel at the same velocity in free space. In the presence of matter, the progress of the wave is affected by its interaction with atoms or molecules, and the velocity depends on both the properties of the medium *and* on the frequency of the wave; it may even (for non-uniform or crystalline materials) depend on the direction of the wave through the medium, and on its plane of polarization.

Refractive index

This is a measure of the extent to which light is slowed down by a medium. For a given frequency (colour) and a given medium, we can measure the refractive index $n = v_0/v$, where v_0 and v are the velocities *in vacuo* and in the medium; since v is normally less than v_0, n will be greater than unity. A given medium has different values of n for different colours. This variation is called *dispersion*. In designing an optical system one needs to know the degree of dispersion because the refraction or bending of light by a lens depends on n, following Snell's Law (formula in Fig. 4). This law applies to the interface between any two transmitting media. Dispersion

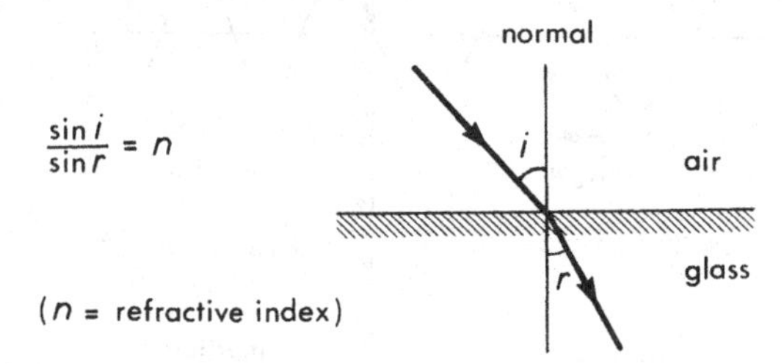

Fig. 4. Snell's Law.

leads to a splitting up of white light into colours and causes chromatic aberration, which we discuss in Chapter 9.

Wavelength

We have so far avoided mention of the wavelength, though the colour of light is commonly expressed in such terms – for instance 546 nanometres (1 nm $= 10^{-9}$ m) for the mercury green line. The omission was deliberate, because although the frequency of a given wave is constant, its wavelength depends on the medium through which it is passing. Fig. 5 illustrates this. S is a source generating field oscillations at a fixed frequency of ν (Greek nu) cycles per second; in one cycle, the electric field passes from its maximum positive value to a negative peak and back again. The wave spreads outwards at velocity v, and we have joined points of maximum (positive) electric field by lines called *wavefronts*. The distance between one wavefront and the next is defined as the wavelength, λ (lambda); it is analogous to the distance between ripples on a pond.

If the velocity decreases on entering another medium, the peaks will be crowded together and the wavelength reduced. That this must be so is seen when we consider what happens at a boundary (Fig. 6); the number of

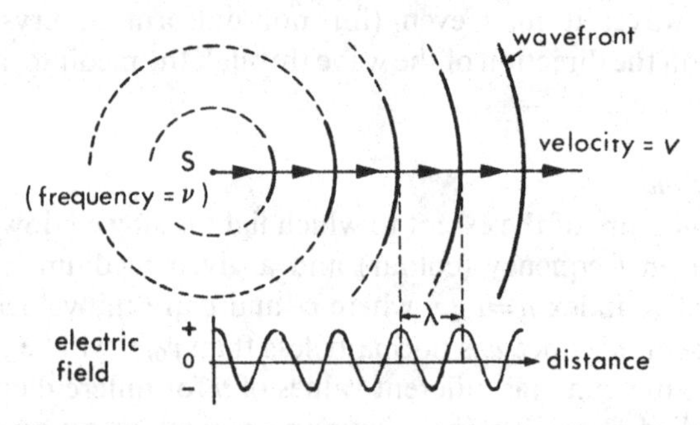

Fig. 5. Wavefronts from a point source.

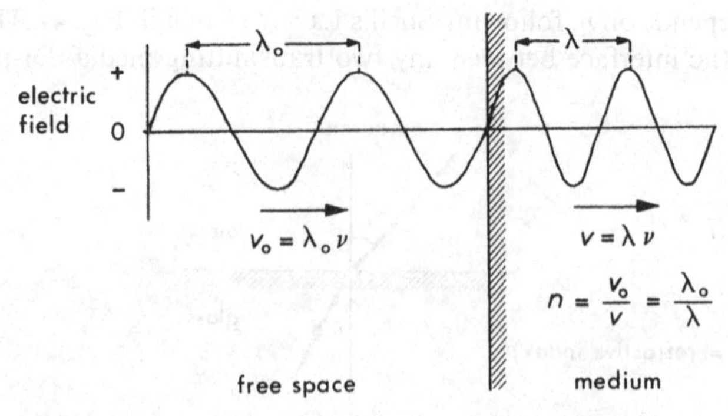

Fig. 6. Change of velocity at a boundary.

peaks arriving per second must equal the number leaving per second from the other side, and the slower they travel the closer together they must be. Since the frequency ν measured by a stationary observer must be equal to the distance travelled per second by the wavefronts divided by the distance between wavefronts, we obtain $\nu = v/\lambda$ and $\nu = v_0/\lambda_0$, from which follow the equations in Fig. 6.

As mentioned earlier, it is customary to specify a colour by its wavelength, but this refers only to the wavelength λ_0 in free space. In air, the value is almost the same – only in liquids or solids does it shorten appreciably. This leads us to an alternative definition of refractive index $n = \lambda_0/\lambda$ (see Fig. 6); λ is normally less than λ_0.

Wave interference

Let us now imagine a stationary observer sitting in a medium as the waves go past. He or she will observe maxima separated by a time $1/\nu$, the *period*. The waves will have a form like those of the 'frozen' waves considered earlier, and the period will be independent of the medium. If a second wave with the same frequency is combined with the first, their net effect will depend on whether or not the positive peaks overlap each other (Fig. 7). E represents the electric field.

If the peaks of two waves overlap exactly (when they are said to be *in phase*), the *amplitude* (maximum field value) obtained by combining waves of amplitudes a_1 and a_2 will be $(a_1 + a_2)$; if the positive peaks of one wave overlap the negative peaks of the other (waves *out of phase*), the resultant has amplitude $(a_1 - a_2)$. *Interference* describes this summing of waves. The intensity (proportional to energy carried – see Appendix) is the square of the resultant amplitude, and is related to the observed brightness. Fig. 7 shows that this can be either greater or less than those of the individual waves.

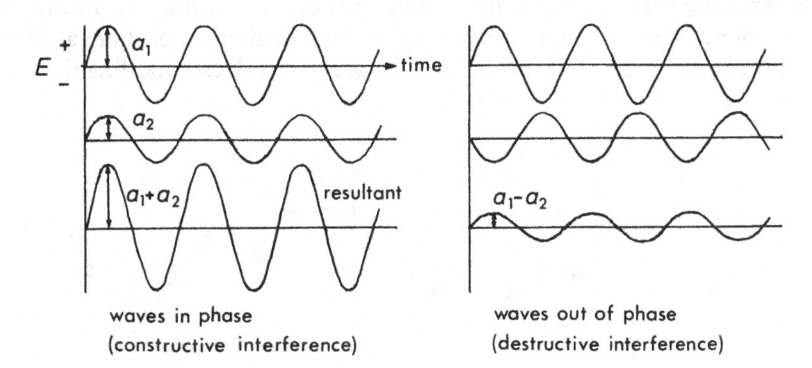

Fig. 7. Interference between waves.

In order to tackle a case intermediate between the two extremes described above we must elaborate the concept of phase. Any wave of the type considered here has a form which is *sinusoidal*, meaning that whatever is varying with time or distance fluctuates like the sine function given in mathematical tables. We define the point reached in the cycle in terms of *phase angle*, θ (theta), derived from the equation $E = a \sin \theta$; here a is the amplitude of the sinusoidal fluctuation in field E. In one complete oscillation, θ covers a range of 360° or 2π radians; it has no geometrical meaning in this context.

Instead of plotting field against time as in Fig. 7 we can plot it against the equivalent phase angle, and for two waves that have the same frequency but do not exactly overlap we can define a *phase difference*, ϕ (phi), that will be independent of where we are on the wave (Fig. 8); it is simply the difference in phase between any pair of corresponding points (e.g. for maximum positive field) on the two waves. Returning to Fig. 7, we see that the waves 'in phase' had a phase difference of zero, while those 'out of phase' had $\phi = 180°$.

In order to work out the brightness of the image when two waves interfere we have to work out the resultant amplitude. To do this when the phase difference is not zero or 180° we need one more concept, that of *vectors*. You may be familiar with these from elementary mechanics, where they are used to work out the resultant of two forces; each force is represented by an arrow whose length varies with the magnitude, while its direction on the paper represents the line of action of the force. Two forces $\mathbf{F}_1$ and $\mathbf{F}_2$ acting upon an object then yield two sides of a parallelogram (Fig. 9), whose diagonal represents in magnitude *and* direction the resultant force $\mathbf{F}_R$. The heavy type indicates that we are now dealing with vectors, and $\mathbf{F}_R$ is said to be the *vector sum* of $\mathbf{F}_1$ and $\mathbf{F}_2$.

To add a pair of waves we can follow exactly the same construction, but the length of each arrow now represents the amplitude, and their relative orientation depends on the phase difference between them. Fig. 9 shows a pair of vectors $\mathbf{a}_1$ and $\mathbf{a}_2$ with a phase difference ϕ, giving a resultant $\mathbf{a}_R$ with a phase difference ϕ_R from the first component. In this case the directions of

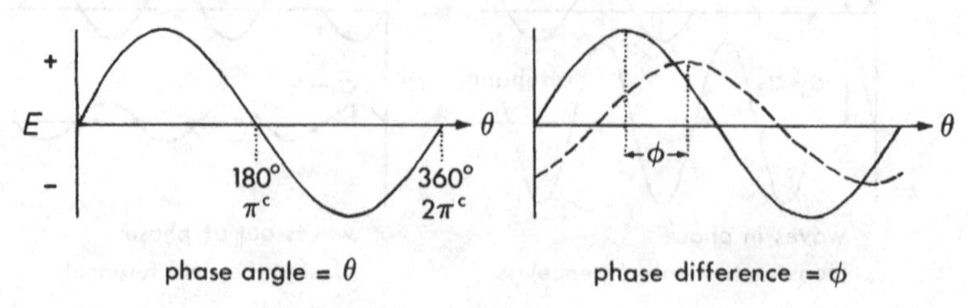

Fig. 8. Phase angle and phase difference.

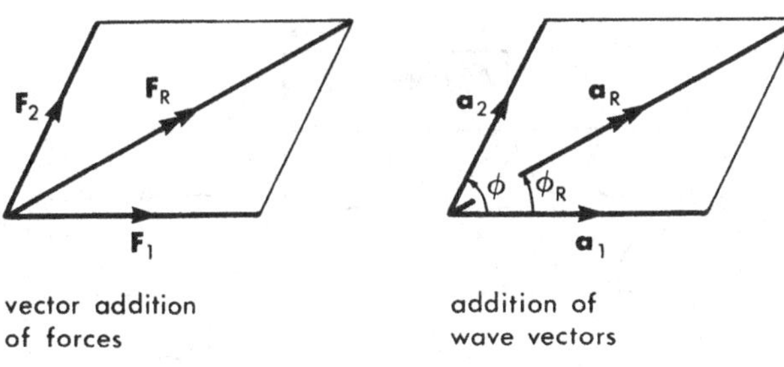

Fig. 9. The use of vectors.

the vectors (unlike those of force vectors) *do not* represent directions in space; the diagram is only a convenient calculating device.

The converse must also hold, that any vector is equivalent to a pair of new vectors making up the sides of a parallelogram, of which the original vector is the diagonal. We shall use this construction in analyzing the phase microscope.

We have assumed in all this that the field vectors of two waves are capable of being added in the simple way described, but this presupposes that the actual orientations of the vectors in space (i.e. the directions in which field fluctuations are observable) are the same for both waves – as they are when two sets of ripples on a pond are combined. Situations where this does not apply form the basis of polarizing and interference microscopy, and we shall consider them in later chapters.

Optical path

In quantitative techniques, such as mass measurement by interference microscopy, we measure the phase difference between a pair of waves, which are generally derived by splitting an incident wave into two components. However, the phase difference may amount to many whole cycles, and the concept of phase angle is inconvenient; it is replaced by that of optical path. The shift can arise either because two waves have followed different paths of unequal length in the same medium, or (more commonly in our applications) because they travel the same distance but the media differ in refractive index. Fig. 10 illustrates this: the two waves travel the same distance, but because one of them is slowed down by the glass it acquires a phase shift relative to the other, *as if* it had travelled a greater distance; the optical paths are said to differ.

The *optical path difference* (o.p.d.) in Fig. 10 clearly depends on both the thickness of the glass block and its refractive index, which as we have seen is related to the wave velocity in the glass. The block contains n times as many wave cycles as the equivalent path in air, and the optical path in the block is defined as nt where t is its thickness. We can, in fact, think of the optical

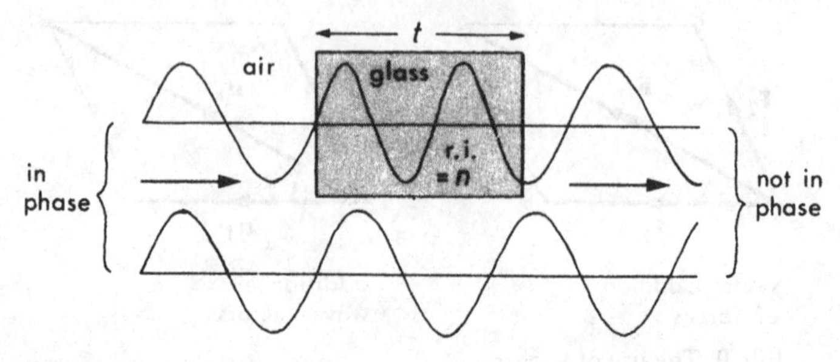

Fig. 10. Optical path difference.

path as *the distance in free space containing the same number of wave cycles as the geometrical path in the medium*; imagine the glass block in Fig. 10 stretched out (with the waves contained in it) to give the same wavelength as in air. The relation between phase difference ϕ and optical path difference Δ (delta) is simply $\phi/2\pi = \Delta/\lambda_0$: they are proportional to each other. In Fig. 10, $\Delta = (n-1)t$ because the refractive index of air is close to unity.

The o.p.d. is also sometimes expressed as a *number of wavelengths* $N = \Delta/\lambda_0$; if N is any whole number, the waves will be in phase and will reinforce each other, because shifting a wave by a whole cycle does not change the positions of the maxima. On the other hand if $N = 0.5$, 1.5, 2.5 etc. the waves will be out of phase, and their amplitudes will subtract from each other.

It is only a small step from the case shown in Fig. 10 to that where the two waves traverse slabs of media which are both different from air and have different refractive indices, say n_1 and n_2. The o.p.d. is then

$$\Delta = (n_1 - n_2)\,t$$

This equation is basic to both polarizing and interference microscopes, in each of which Δ is measured with a calibrated device called a *compensator* (of which more later). From Δ we can then either deduce $(n_1 - n_2)$ if t is known, or vice versa. In the interference microscope Δ is directly related to the mass per unit area of the object.

Questions

1. What happens to monochromatic light passing through a stained object, and what is permanently changed by the process?
2. What property do waves and moving particles have in common?
3. How do the field fluctuations in light waves differ from the pressure variations in a sound wave?
4. What happens to the wavelength of light if its velocity decreases on crossing a boundary between two media?

5. When a wave enters a different medium, what happens to the *rate* of fluctuations measured by a stationary observer?
6. In the case described above, what happens to the colour?
7. Two waves differ in phase so that the maxima of one overlap the zeros of the other. What is their phase difference in degrees?

3 The limits of resolution

We are nearly in a position to understand the theories which explain why there is a limit to a microscope's *resolution* (ability to distinguish closely-spaced objects). First, however, we must consider one more aspect of interference, and we must understand something about *diffraction.*

When light is collected from two independent sources, or from two regions of an extended source like a discharge lamp, we cannot say whether the two sets of waves are in phase or not, because the relative phases are changing all the time; this is because (as mentioned in Chapter 2) each source gives out randomly-timed quanta of radiation. The sources are said to be *incoherent*. However, if light from one source is optically split into two component beams (as at a half-silvered mirror, which transmits a proportion of the light), then the components are *coherent* and can produce interference phenomena if they are subsequently brought together.* For incoherent sources, the energy received is proportional to the sum of the intensities, but when the sources are coherent we must first work out the amplitude resulting from the interference, and square that to obtain the intensity (the mathematical reasons for this are given in the Appendix).

Diffraction is a phenomenon important in microscopy, but there are few everyday examples of it. The simplest way of demonstrating it is to observe a distant light source (such as a naked bulb) through a fine handkerchief or a woven umbrella held close to one eye; you should then see a pattern of spots and streaks of light whose orientation is determined by the weave of the material. Light from the source has not only travelled in straight lines to your eye, but some of it has been diverted by the small holes in the fabric to blur the image of the source. We shall see that the same kind of effect limits the resolution of the microscope, but in this case the diffracting aperture is that of the objective.

To work out the effects of diffraction at a single aperture we imagine it

* According to quantum mechanics, we have to talk not of each photon being split up, but of there being a certain probability that it will follow one path rather than the other. See Jenkins and White (1976) (listed in Chapter 11) for a discussion of this problem.

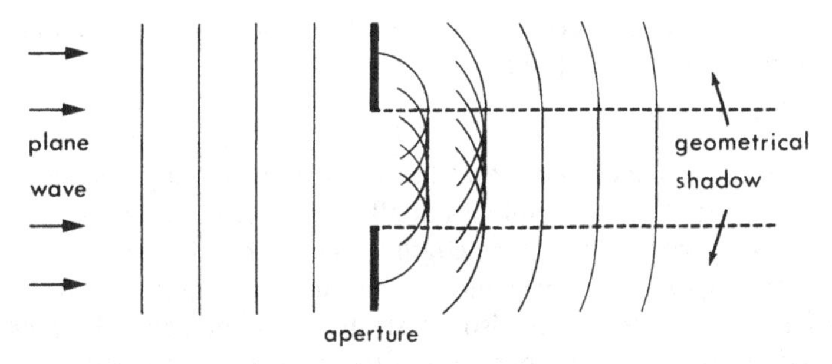

Fig. 11. Huygens' wavelets.

illuminated by a *plane wave*, as from a distant source or a beam of light made parallel by a lens; the wavefronts are straight lines. At an aperture (Fig. 11) we imagine each point on a wavefront acting as a new source of light, giving out spherical wavefronts called *Huygens' wavelets* (after the seventeenth-century scientist who introduced the idea). The new imaginary sources are all coherent with each other, and the form of the wavefront after the aperture is calculated by adding their contributions together. By doing this we find that some light is propagated into the geometrical shadow of the aperture; the smaller the aperture, the more noticeable this will be.

The mathematics of a large aperture is complex, but a simple example (devised by Thomas Young in the early nineteenth century) will illustrate the principle; a class demonstration of it is described in Chapter 11. Wavefronts from a narrow slit aperture in front of a source S (Fig. 12) illuminate two more slits A and B. These act as coherent sources, and the waves from them interfere to produce a regular set of *interference fringes* on the screen. Whether the intensity at a point P is a maximum, or zero, depends on the path difference (BP – AP); when this is a whole number of wavelengths, the wavefronts originating from A and B will exactly coincide when they reach P, and constructive interference will give a bright fringe.

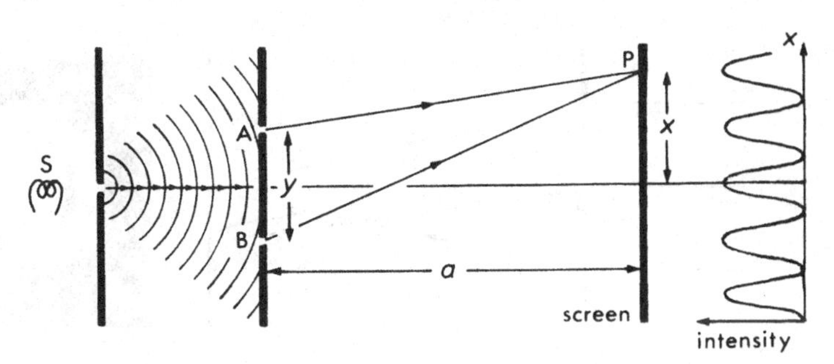

Fig. 12. Young's experiment.

When x and y are small compared with a, the condition for a bright fringe is shown in the Appendix to be

$$x = N\lambda a/y$$

where N is an integer. Note that the separation of fringes is $\lambda a/y$, which depends on λ; in this example, as in all diffraction, the *scale* of the fringe pattern increases with the wavelength. Thus, with red light the fringes will be further apart than with green, because red light has the greater wavelength. The equation also illustrates another general property of diffraction patterns: the separation of fringes *decreases* as the scale of the diffracting structure (measured here by the aperture separation y) gets larger.

Now consider what happens when a lens is illuminated by a beam of parallel light (Fig. 13). Because of diffraction at the aperture of the lens the light does not focus to a point, but forms a diffraction pattern consisting of an intense central maximum flanked by weak secondary maxima. (In this and other diagrams the scale of the pattern has been greatly exaggerated.) The effect was first noted by astronomers, who found that the image of a star had a finite size which depended on the diameter of the telescope's field lens. In 1834 Airy worked out that for diffraction at a circular aperture of diameter d (Fig. 13) the *angular* position of the first minimum (counting from the centre) is given (see Appendix) by

$$\sin \alpha = 1.22\lambda/d$$

This pattern is called the *Airy disc*. You can easily observe it by holding close to one eye a card in which there is a very small pinhole, while you look towards a bright point source such as a microscope illuminating bulb, at a distance of several metres (try holes of different diameters).

Exactly the same kind of pattern is formed when a point object is imaged by a microscope; the diffracting aperture is in this case that of the front lens of the objective. The effect only becomes noticeable at the highest magnifications, when every speck of dirt gives a weak diffraction pattern.

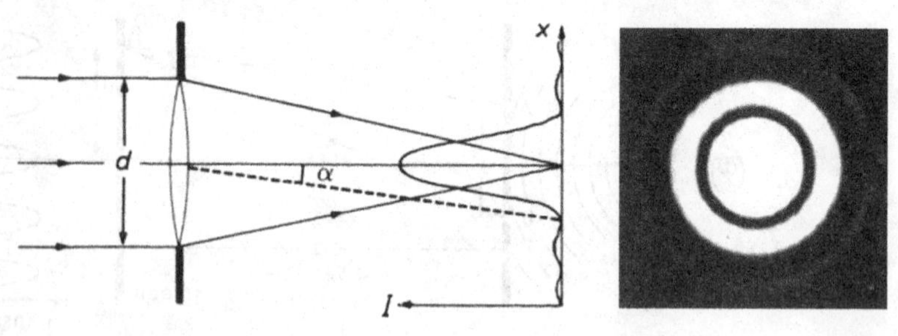

Fig. 13. Diffraction at the aperture of a lens: the Airy disc.

The Airy disc is demonstrated most clearly when the object is a metal-coated slide containing minute pinholes. We shall see in Chapter 9 that this forms the basis of a method for testing the quality of an objective.

The Rayleigh criterion

There is no lower limit to the size of the object that can be detected in isolation, given adequate illumination. However, if a single point object gives an image like Fig. 13, then clearly a second point close to the first will give a similar diffraction pattern that overlaps the first. There comes a point where the overlap is so bad that one cannot say whether there are two objects or only one. The definition of this point is somewhat arbitrary, but Rayleigh suggested a criterion which applies well to most situations: that two patterns will be distinguished when the central maximum of one pattern lies over the first minimum of the other. The intensity (Fig. 14) then falls to 80% of the maximum between the two peaks.

The application to a microscope is shown in Fig. 15. We have to allow for the fact that if an immersion medium is used between the slide and the objective, the image-forming ray passing through the centre of the lens is bent; in the medium, the ray makes an angle β with the axis which is smaller than α, the angle in air. A given size of Airy disc therefore corresponds to a smaller object-separation y_R than in air. The full mathematical treatment (see Appendix) shows that for satisfaction of the Rayleigh criterion

$$y_R = 0.61\ \lambda_0/n \sin\theta_0$$

Here λ_0 is the free-space wavelength, n the refractive index of the immersion medium, and θ_0 half the angle subtended by the lens at the object.

We can see at once why the use of immersion oil improves the resolution. The product $n \sin\theta_0$ is fixed by the design of the objective, and is called the *numerical aperture* (N.A.); it is this rather than the magnification that really determines the ability to image fine detail. Clearly the N.A. should be as large as possible but there are practical limits to both n and θ_0, so that the N.A. rarely exceeds about 1.3. This sets a limit to the resolution of about

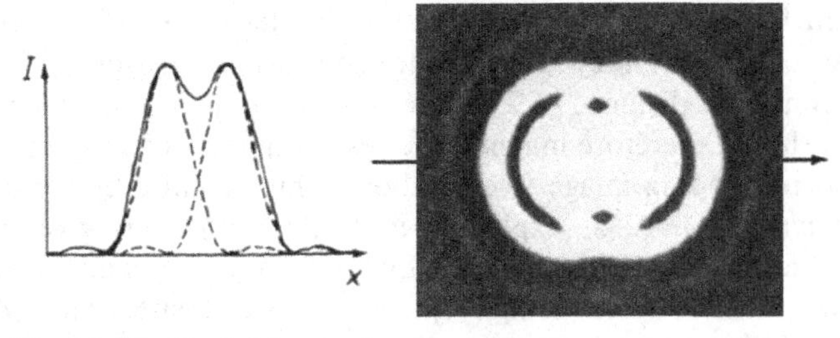

Fig. 14. The Rayleigh criterion for resolution.

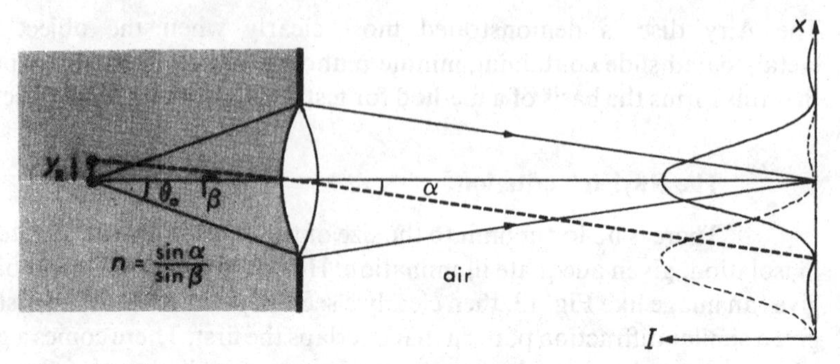

Fig. 15. The resolution limit of a microscope.

half a wavelength, when using a high-power objective; lower-power objectives have smaller N.A.s and the limit is greater. In principle we should therefore use the shortest possible wavelength, and before the advent of the electron microscope there was considerable research on ultraviolet microscopy. However, the use of ultraviolet light poses technical problems that nowadays make it seldom worthwhile. For green light, the Rayleigh criterion means that the resolution limit is about 0.25 μm or 250 nm.

Limits to magnification

Having established a theoretical limit to the resolving power, we need to ask how much *total* magnification is useful in a microscope. There is clearly no advantage in magnifying the image to the point where it is confused by all the secondary diffraction maxima (Fig. 13); such *empty magnification* will show only spurious detail that could be very misleading. Examples of this are still to be seen in scientific journals, and electron microscopy is subject to the same abuse.

The best compromise is to match the image to the resolving power of the detector – the human eye or a photographic film. The eye, too, is limited in resolution by its aperture (the iris), and the detectors in the retina are spaced accordingly. The eye's limit y_R for comfortable viewing is about 200 μm (0.2 mm) at a distance of 250 mm. A microscope using a high-power objective should therefore magnify the resolving limit of 250 nm so as to give a corresponding image size of about 200 μm (an eyepiece actually forms a *virtual image* as explained in Chapter 1). This gives a total magnification of 800, so we can see why for an objective magnification of × 100 there is little point in using eyepieces of power greater than × 8. With fine-grained photographic film even this may be too much; the greater the magnification, the longer will be the exposure time. Photomicrography is considered in greater detail in Chapter 10.

Comparisons with electron microscopy

The same theory can be used to predict the resolution limit of an electron microscope. Moving electrons have an equivalent wavelength which is given approximately (in nm) by $\lambda = \sqrt{1.5/V}$ where V is the accelerating voltage; thus for $V = 60$ kV, λ is 0.005 nm, which is 100 000 times smaller than the wavelength of green light (see the book edited by Glauert (1974), listed in Chapter 11). However, to reduce the aberrations (see Chapter 9), the numerical apertures of the electron lenses have to be very much smaller than in a light microscope. This reduces the advantage given by the smaller wavelength, so that the theoretical limit is about 0.2 nm. For biological material, problems with specimen preparation raise the resolution limit to about 1 nm, or about 250 times smaller than in light microscopy.

The small numerical aperture of an electron microscope gives it a very large depth of field (Chapter 1), so that normally the whole specimen is sharply imaged. This means, however, that one cannot get an idea of three-dimensional structure by focusing at different levels within a thick specimen; the specimen must be tilted instead.

Questions

1. In the example given earlier of a pinhole held up to the eye, what determines the diameter of the Airy disc – the size of the pinhole or that of the eye pupil?
2. In the same example, white light gives multicoloured rings round a white central maximum. Why is this?
3. If green light is replaced by blue, how does the scale of a diffraction pattern change?
4. Can the resolution of an image be improved by recording it on fine-grained photographic film, instead of viewing it by eye?

4 The Abbe approach

We looked at the resolution limit first from the viewpoint of the Rayleigh criterion, because this is a relatively simple analysis that gives roughly the right answer. However, it has two defects. Firstly it assumes that the point objects considered behave like self-luminous incoherent sources. This is applicable to stars viewed through a telescope, but not strictly correct in a microscope – the illumination does in fact have some degree of coherence across the object field, depending on the separation of the points considered and the design of the microscope. The second problem is that it is not clear how we can predict the effects of diffraction on a complex object – which details are affected, and how?

A more satisfactory approach is that of Abbe, who in the nineteenth century initiated the idea (widely misunderstood at first) that Fourier analysis can be used to predict the form of the image produced by an objective of limited aperture. We shall use this approach firstly to show how an alternative expression for the resolution limit can be derived; however, its main usefulness lies in its ability to explain how the phase microscope works, and we shall deal with that in Chapter 5. There is also a 'spin-off' in that the same approach sheds light on the operation of both optical and X-ray diffraction.

Fourier analysis

Fourier's theorem (see Appendix) shows that even an irregular curve, such as a plot of the brightness along a line crossing the field of view of a microscope, can be regarded as the result of adding a large number of *regularly*-fluctuating amplitude variations of differing repeats – more precisely, the sum of a series of sinusoidal variations whose periods are sub-multiples of the largest periodicity or repeat distance present. The amplitudes of these Fourier components generally get smaller as the periodicity decreases, but the short periodicities contribute to the sharp edges and fine detail of the image profile.

Fig. 16 shows this for a repeating rectangular wave. There is a steady

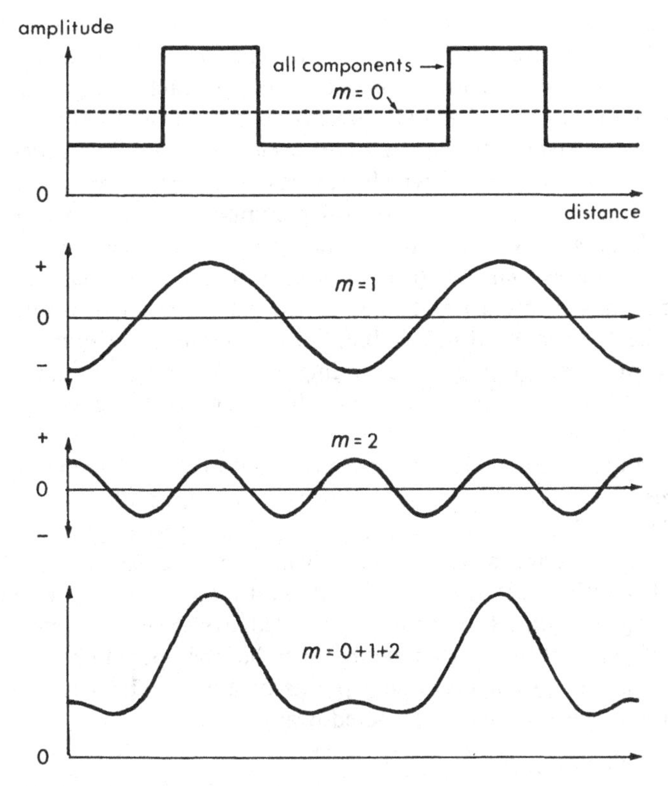

Fig. 16. Fourier analysis of a rectangular wave.

background labelled $m=0$ and called the *zero-order* component, and a large sinusoidal *first order* ($m=1$) with the same repeat as the original wave. The next component ($m=2$) has half the basic periodicity; other components (not shown) have repeats of 1/3, 1/4, 1/5 etc. of the basic value. If we add the components for $m=0$, 1 and 2 we get a profile with some resemblance to the original wave, but with the sharp edges missing and some distortion elsewhere. As higher-order components are added, the sum will look more and more like the original. However, any cutoff of higher-order components must introduce some distortion. We shall see that this enables us to predict how the quality of a microscope image is affected by the finite aperture of the objective.

The application to image formation

We shall use a simple but still rigorous approach which is different to that found in most textbooks. Let the rectangular-wave profile of Fig. 16 represent the distribution of amplitude produced by an absorbing object which consists of an array of alternating light and dark strips. Imagine the zero-order component to be divided up into parts which are then added to the various sinusoidal components so as to raise each one just above the zero-amplitude axis; this will give positive amplitude at all points, instead of a phase reversal between each half cycle. (There will in general be a residual $m=0$ contribution, which may be either positive or negative.) We can now imagine each non-zero-order component as representing the amplitude passed by a 'grating' of the appropriate repeat, in which the absorbance varies *sinusoidally* across it as in Fig. 17; the 'bars' of such a grating do not have sharply-defined edges. The rectangular profile of Fig. 16 can then be produced by combining the separate contributions from a number of these gratings and adding or subtracting an appropriate background component. (This is not, however, the same as laying one grating over another.)

Now, any repeating structure gives rise to diffraction which is specially reinforced in certain directions. In the case of a sinusoidal grating of the kind described above the energy is divided between an undeviated zero-order beam and a pair of beams which make equal angles with the central beam (Fig. 17); there are no other diffracted beams. We can work out the angle of diffraction from our knowledge of the laws of interference. For reinforcement, the angle must be such that the optical path difference between rays from corresponding points of adjacent bars is one whole wavelength; hence for each diffracted beam

$$y \sin \theta = \lambda$$

The situation is, in a sense, the reverse of that in Young's experiment (Fig. 12), where diffraction from a pair of sources gave rise to a repeating diffraction pattern: here we have a repeating grating and only two diffracted beams. If we imagine such a grating illuminated by parallel light

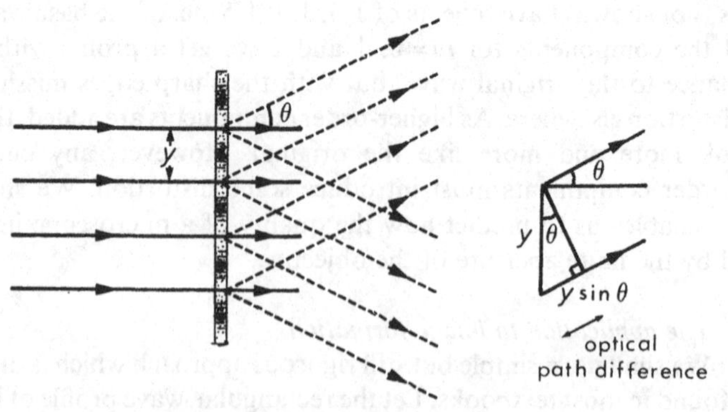

Fig. 17. Diffraction from a sinusoidal grating.

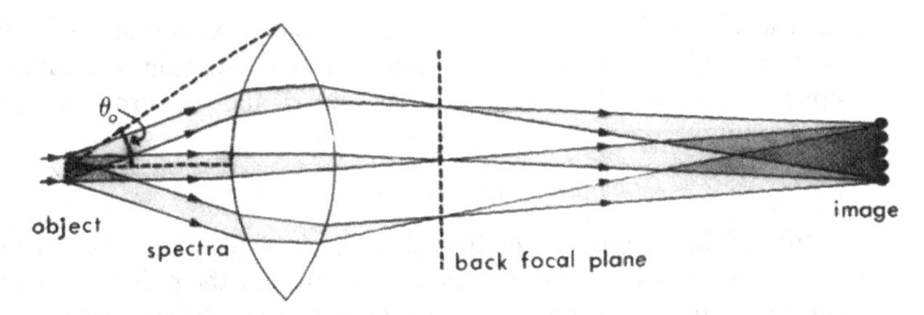

Fig. 18. The Abbe treatment of image formation.

and imaged by a microscope objective (Fig. 18), each of the diffracted beams (often called 'spectra' because of their appearance when white light is used) will be brought to a focus behind the lens at the back focal plane (observable with a phase telescope or swing-out lens). The beams will then diverge as they travel towards the image plane, where they interfere with each other to give a distribution of amplitude which is a magnified version of that produced by the object.

In the case of an irregular object the only difference is that many more spectra, from the different 'component gratings', make up the image. As in Young's experiment, interference is not limited to one plane; the beams interfere wherever they overlap, but away from the image plane there is no longer a meaningful reconstruction of the object: the image is 'out of focus'.

The resolution limit

In the case of an irregular object the smaller component spacings will give larger values of the diffraction angle θ. Beyond a certain value of θ (equal to θ_0 in Fig. 18) they will miss the objective aperture altogether and cannot be collected to reconstruct the image. There will then be distortion of the kind illustrated in Fig. 16, where higher-order components were left out; the fine detail will be lost. To form a meaningful image, we must collect the zero-order and at least the first-order spectra that arise from the largest spacing present. Applying our diffraction equation to the limiting angle θ_0 in Fig. 18, we then obtain

$$y_R = \lambda/\sin\theta_0 = \lambda_0/n\sin\theta_0 = \lambda_0/(\mathrm{N.A.})$$

The refractive index comes in because we are concerned here with *optical* paths (Chapter 2); N.A. means numerical aperture (Chapter 3).

This is almost the same as the expression derived from the Rayleigh criterion, except that in that case there was a multiplying factor of 0.61, suggesting a smaller resolution limit. However, we have neglected the fact that parallel-light illumination is not normally used; the condenser gives instead a cone of illumination so that some of it strikes the grating obliquely (Fig. 19). If we assume that to resolve a given spacing it is sufficient to collect only *one* of the two first-order spectra and the zero-order one, the expression becomes

$$y_R = \lambda_0/2n \sin \theta_0 = 0.5\, \lambda_0/(\text{N.A.})$$

Fig. 19 illustrates one of the advantages of using a condenser whose numerical aperture is at least as great as that of the objective: it actually improves the resolution because of the increased obliquity of some of the illumination. Abbe suggested an empirical rule for intermediate cases,

$$y_R = \lambda/((\text{N.A.})_{\text{obj}} + (\text{N.A.})_{\text{cond}})$$

For both N.A.s equal to 1.3 this gives a limit of about 0.2 μm in green light. Even the Abbe treatment is not perfect; we have assumed (in complete contrast to the Rayleigh approach) that the illumination is coherent. This is not strictly true, and the real situation (whose mathematical treatment is complicated) lies somewhere between the two extremes. It can be shown that, although the condenser aperture has some effect on the resolution, it is not as important as the Abbe treatment suggests. However, there are other sound reasons (dealt with elsewhere in this book) for having a large condenser aperture. Where the Abbe approach is particularly useful is in its prediction that spurious detail will appear in the image when higher-order spectra are not collected. Special caution is therefore needed when the size of an object approaches the resolution limit. Kits to illustrate this and other aspects of the Abbe theory are available from Zeiss (West Germany).

For anyone who is still unconvinced of the need to use an immersion liquid for high magnifications, the Abbe theory offers an alternative way of explaining it. The diffracted spectra from an object mounted in liquid (Fig. 20) will be refracted at a boundary with air, and in the limit will miss the

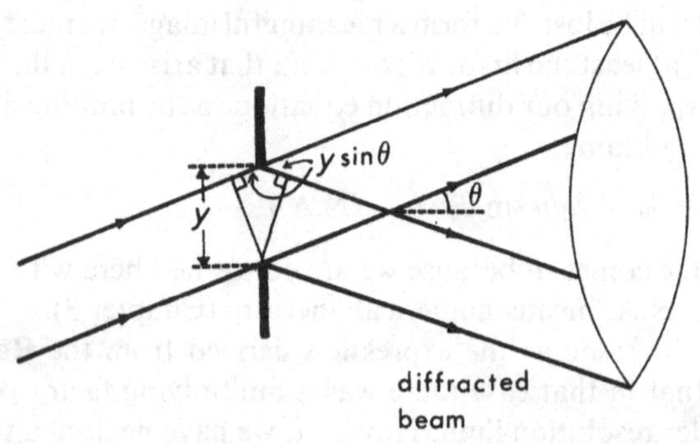

Fig. 19. Oblique illumination of a diffracting object.

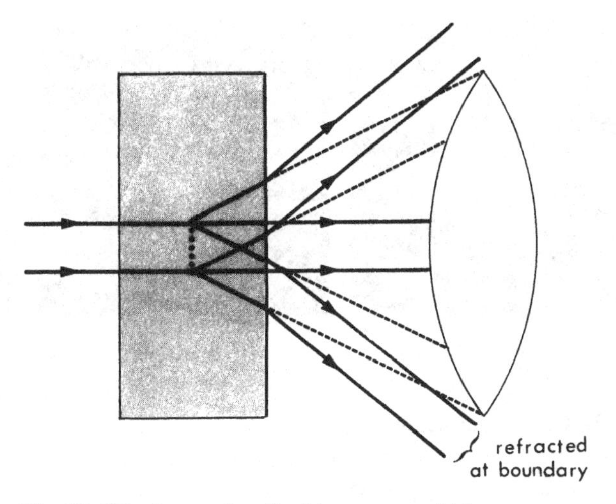

Fig. 20. Why immersion liquids are a good idea.

objective so that the image is impaired or even lost. With an immersion liquid between object and lens the effect is much reduced.

Optical and X-ray diffraction

Optical diffraction is increasingly popular as a means of 'cleaning up' electron micrographs of repeating structures, and showing up hidden periodicities. The method is exactly the same as that shown in Fig. 18, with a suitable area of the micrograph acting as the object for a large lens. Diffracted beams in the back focal plane are then recorded on a photographic plate, and regular repeats in the structure show up as strong spots superimposed on a 'noisy' background. The repeat spacings can be measured directly from these spots, or a mask can be introduced to remove the noise and a 'filtered' reconstruction observed in the image plane. Fig. 21 shows the technique applied to an array of thin filaments from muscle. One can also reconstruct separately the 'front' and 'back' of a three-dimensional object.

There is, of course, a risk that by over-zealous filtering one can introduce a non-existent periodicity, so the technique has to be used with care. A more sophisticated application of optical diffraction involves using a series of micrographs taken from different angles to reconstruct a three-dimensional model of an object, but this is strictly one for the specialists.

In X-ray diffraction one is limited by having no lenses that will focus X-rays; otherwise one could simply form an image of a crystal as in a microscope. All that one can do is to record the diffracted beams leaving a

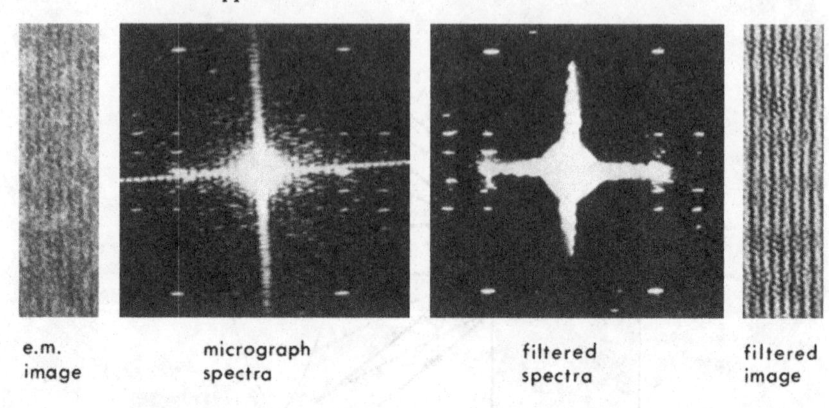

Fig. 21. Optical diffraction.

crystal; if the crystalline array is very regular there are many well-defined spots (Fig. 22). One then has to perform mathematically the reconstruction (in three dimensions here) that a microscope does optically, and there are many technical problems that need not concern us here. However, when it is successful the method can determine the positions of the atoms in a molecule to within a small fraction of the wavelength of the X-rays, which is commonly about 0.15 nm. The 'resolution limit' is much smaller than the wavelength because information about atomic structure can be utilized.

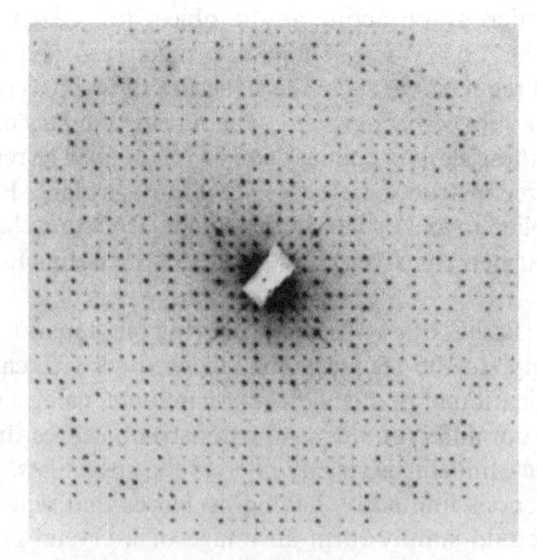

Fig. 22. X-ray diffraction from a protein crystal.

Questions *(quite hard – do not be discouraged by failure)*

1. If the first-order spectra from a regular grating are masked off in the back focal plane of a microscope, and only the zero- and second-order spectra are passed, what will happen to the image?
2. What will happen if both first-order beams are collected but the zero order is masked off?
3. A diatom has an internal periodicity of 0.2 μm which can just be seen with a $\times 100$ objective. Could the same structure be seen by appropriate magnification of the image from a $\times 40$ objective?
4. Why is X-ray diffraction sometimes called a branch of microscopy?

5 Bright-field, phase and dark-field microscopy

We saw in Chapter 4 how, with an absorbing object, diffraction spectra are produced as if the object consisted of a set of gratings of different periodicities, and that the spectra can be recombined to construct an image. However, to many of Abbe's contemporaries this approach seemed inadequate to explain the working of the various empirical methods by which transparent objects were made visible. How could a transparent object give spectra? Why did the contrast improve when a microscope was slightly defocused, or oblique illumination used? For some time, many people rejected the Abbe theory as irrelevant to practical microscopy.

It was Zernicke in the 1930s who clarified the situation by his invention of the phase-contrast microscope. Ironically, he had great difficulty in persuading some disciples of Abbe that a new approach would be fruitful. Zernicke's contribution was to point out that with interference phenomena, the phase as well as the amplitude of each ray must be taken into account. A 'phase grating' made by forming grooves in a transparent sheet does in fact produce spectra just like an amplitude grating with parallel absorbing bars. In a perfectly-focused microscope of high numerical aperture, the phase-grating spectra combine to produce an image which is invisible to the eye, because the eye is not sensitive to variations of phase. The same applies to an image recorded on film.

However, any disturbance in the distribution of spectra will upset the balance, and some variations of amplitude (i.e. contrast) will be introduced. Since the spectra are converging at various angles to the image plane (see Fig. 18 on p. 27), *changing the focus* (moving away from the image plane) will cause their relative phases to change and produce an amplitude variation; a given shift of the plane of observation will change the optical path more for an obliquely-incident spectrum than it will for the central zero-order beam. Even more contrast is introduced by *closing down the aperture stop*, since this removes the highest-order spectra altogether. In dark-field microscopy (considered in more detail later) the zero-order beam representing the background illumination is removed. Fig. 23 compares the

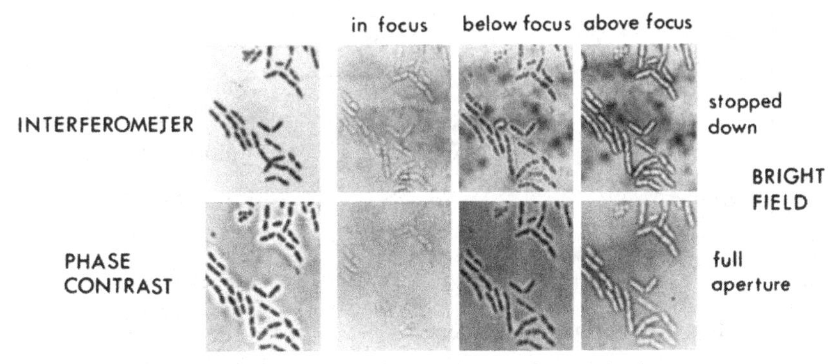

Fig. 23. A slide of bacteria observed by various methods.

images obtained by various methods, including phase contrast and the interferometer microscope described in Chapter 8.

Phase contrast

The basic principle of the method is that the zero-order beam, which follows the direction of the incident light (Chapter 4), has its phase shifted inside the microscope before being allowed to recombine with the other spectra to form an image. The phase shifting is done by introducing into the back focal plane of the objective (where the spectra can be separated) a plate whose effective thickness is altered in the region crossed by the zero-order rays. This leads, in a way that we shall discuss in detail later, to a change of light intensity in the image of a specimen relative to its background. The effect is analogous to a change from a case where two waves are in phase (when they reinforce each other) to one where they are out of phase, when they interfere destructively and give a smaller amplitude (see Fig. 7 on p. 13).

In order to separate the spectra in the back focal plane we must restrict the illumination in some way, because with a full cone of illumination the various zero-order beams for different angles of illumination would completely fill the back focal plane. By simply closing down the aperture stop we should lose a lot of light and also sacrifice some of the resolution. In the design most commonly used now, the aperture stop is replaced by a plate in which there is a clear ring or *annulus* (Fig. 24). The undeviated (zero-order) light is then imaged in the back focal plane of the objective as a bright ring, while diffracted spectra pass mostly through other regions of the back focal plane; they can be seen when viewed through a phase telescope as a pattern of spots and streaks of light against the dark background.

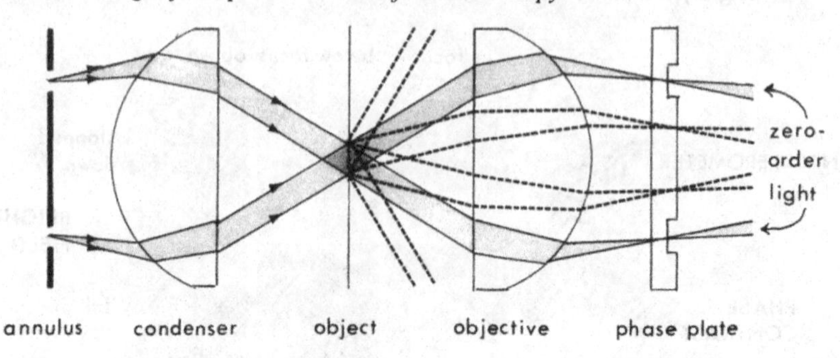

Fig. 24. The phase microscope.

In the back focal plane there is a *phase plate* consisting of a transparent disc in which there is an annulus having (in most cases) a smaller optical path than the rest of the plate. This is arranged to overlap the bright zero-order ring so that its light is *less* retarded (see Fig. 10 on p. 16). Thus it is necessary to match the sizes of the two rings, and to centre one ring on the other (by moving the condenser annulus) before use. In practice the ring in the phase plate is also made slightly darker than its surroundings, for reasons discussed later.

Anyone who has used a phase microscope will have noticed two peculiarities. Firstly, some thick objects can show less contrast than thinner ones, and may even reverse contrast so that they are brighter than the background. Secondly, every object is surrounded by a diffuse halo of light which does not represent real structure. To understand all this we can use the knowledge of vectors acquired in Chapter 2; the original explanations were due to Zernicke.

Diagram 1 of Fig. 25 shows the situation in bright-field microscopy with an absorbing object; the image vector is represented by an arrow whose length is less than that representing the background. The difference between the two is shown by a third vector, out of phase with the others, representing the absorbed light; the image vector is equivalent to the vector sum (taking account of phase) of the other two.

With a transparent object there is no change in amplitude but there is a phase shift relative to the background, indicated in Fig. 25, diagram 2 by the angle ϕ. If the object is made thicker, the image vector will move clockwise round the circle without changing in length, and the object will remain invisible to the eye. The third vector in the diagram represents all the diffracted light, whose combination with the undeviated (zero-order) light produces the image. We have not gained or lost any energy, because the image vector has the same length as that representing the background. As before, the image vector is equivalent to the vector sum of the other two.

Now let use imagine the zero-order vector shifted in phase by 90°, as with a phase plate of the kind described earlier. Fig. 25, diagram 3 shows that this shifted vector, when combined with the unshifted vector for diffracted light, now gives a resultant whose amplitude is reduced. The background remains unchanged because it generates no diffracted light.

In order to work out what happens as ϕ increases, it helps to draw the

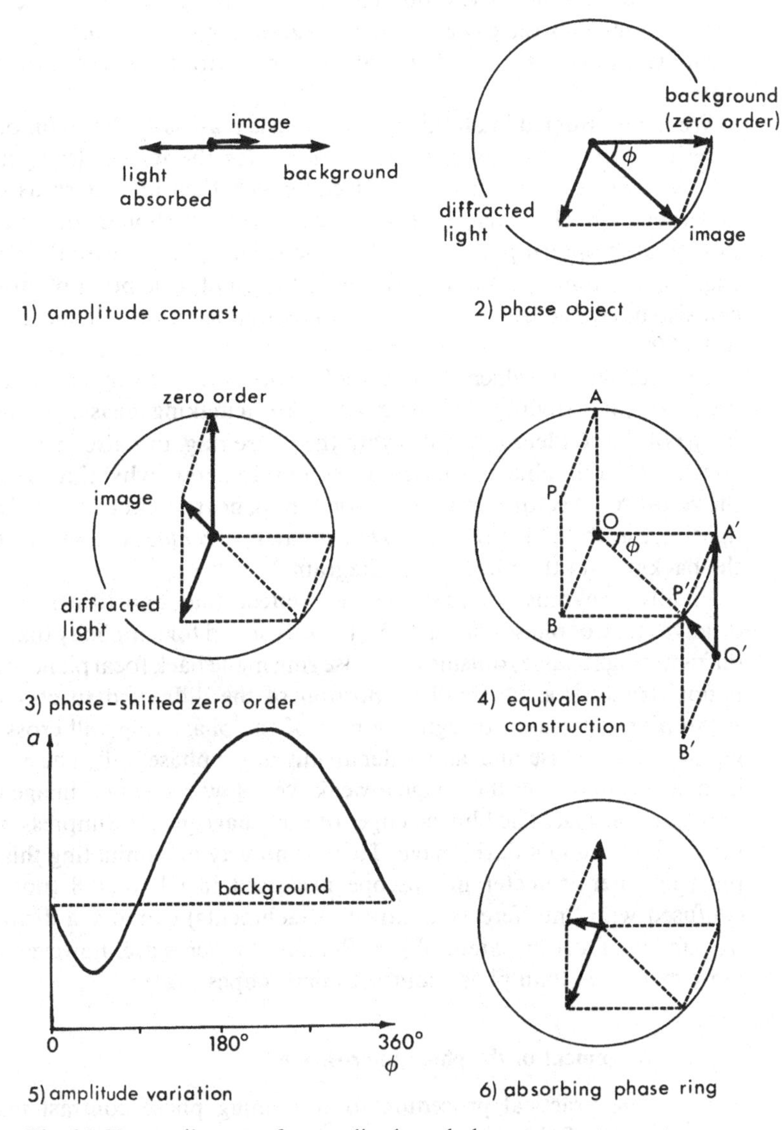

Fig. 25. Vector diagrams for amplitude and phase contrast.

alternative construction in Fig. 25, diagram 4. All that we have done is to move the origin from 0 to 0′ so that the zero order is now **0′A′** instead of **0A**. **A′P′** has the same length as **0B** so that we can draw another line **0′B′** to complete the parallelogram, and its diagonal **0′P′** now represents the image vector instead of **0P**. As ϕ increases, the point P will move round the circle, and we can see that the image vector will increase until it actually exceeds **0′A′** in length. As P′ moves round towards A′ the vector will shorten again until it completes the cycle. Fig. 25, diagram 5 plots this out in graphical form: we can now see why the contrast varies with thickness in the way it does.

We have assumed in all this that the phase plate is of the common kind described, with a ring of reduced thickness for the zero-order light. This gives *positive contrast* for thin objects, which then look dark as if they absorbed light. However, there is no reason why one should not have a ring of increased optical path rather than the reverse; in this case the curve in Fig. 25, diagram 5 is simply reversed left to right. The point of crossover can also be shifted by changing the amount of phase shift – it does not have to be 90°.

One refinement which is commonly employed is to make the ring in the phase plate slightly absorbing. Apart from making it easier to align the image of the condenser annulus with the phase ring, this also improves the contrast for thin objects. Diagram 6 of Fig. 25 shows why: if the length of the zero-order vector is reduced to something nearer that of the diffracted-light vector, the image vector shows a greater *percentage* change relative to the background than in Fig. 25, diagram 3.

Finally, we must discuss the halo effect that constitutes a minor disadvantage of phase contrast. We have assumed for simplicity that all the diffracted light passes outside the phase ring in the back focal plane, but this is not strictly true. A small proportion of the diffracted spectra whose zero-order rays pass through one part of the phase ring will cross other regions of the phase ring, and suffer an unwanted phase shift. The net effect is to superimpose on the image a weak, very low-resolution image of the opposite contrast. The blurred edges of this image give the impression of a bright halo round a dark image. There is no way of eliminating this; only the true interferometer microscope discussed in Chapter 8 (not to be confused with 'interference contrast' attachments) can give a distortion-free image of a transparent object. Such instruments are, however, much more expensive than phase-contrast microscopes.

Alignment of the phase microscope

The practical procedure for obtaining phase contrast may be summarized as follows:

1) Having chosen a phase objective of the desired magnification, select the corresponding condenser annulus; there should be a distinguishing mark on the objective to indicate which annulus to use.
2) Focus on an object and obtain Köhler illumination (Chapter 1) by adjusting the condenser focus and centring controls to give a sharp image of the field stop.
3) Insert a phase lens, or replace the eyepiece by a phase telescope, so as to view the back focal plane of the objective; for this adjustment it is best not to have a thick object in the field. A bright ring will be seen which is an image of the condenser annulus, superimposed on a dark ring which is the phase ring inside the objective. Focus the telescope to obtain a sharp image, and then adjust the annulus centring controls (embodied in the condenser) so that the two rings coincide. The bright ring should appear evenly illuminated.
4) Remove phase lens or telescope and check the earlier adjustments. Remember when changing objectives that a realignment may be needed if the annulus is changed.

Dark-field microscopy

Although this is one of the oldest kinds of microscopy it still finds application to very small objects that are barely visible by other methods. Bacterial flagella, which are only 20 nm in diameter, can be followed in motion by using a very intense source with dark-field illumination; although the diameter is much less than the wavelength of light, the flagella are relatively long and they scatter enough light for the profile to be seen. Nerve cells in culture, which produce very thin processes, are also seen well by this method. Fig. 26 shows some bacteria; it will be seen by comparing

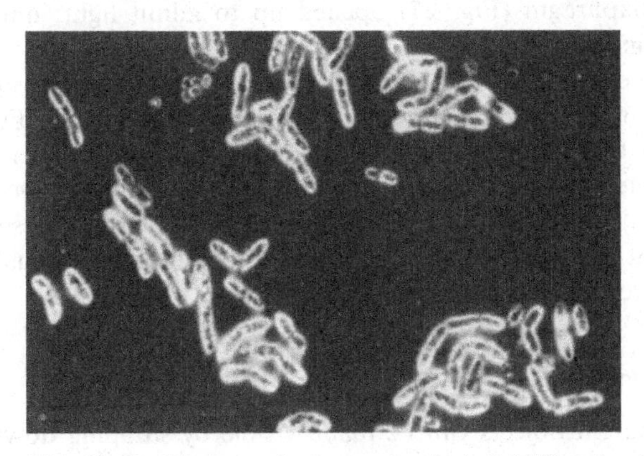

Fig. 26. *Bacillus megatherium* under dark-field illumination.

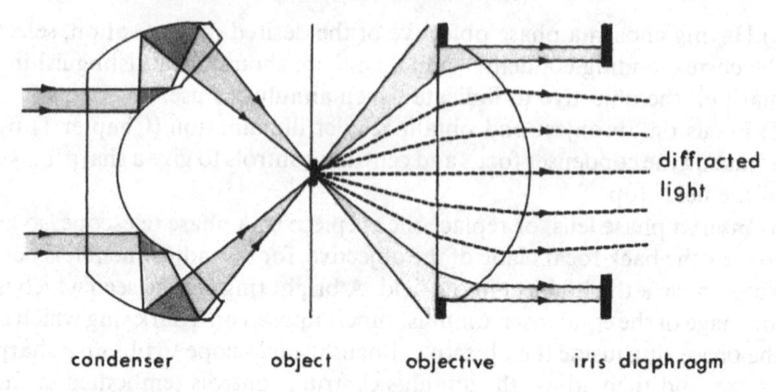

Fig. 27. Dark-field illumination.

with Fig. 23 that only the edges show up clearly, where the refractive index is changing most rapidly. For the reasons explained in Chapter 4 (see answer to question 2), there will inevitably be distortion of the fine detail in the image.

The optical arrangement for maximum brightness at high magnification is shown diagrammatically in Fig. 27; a fainter image can also be obtained with a normal condenser fitted with an annular stop, as in phase contrast. The system produces a hollow cone of illumination, but here the undeviated light misses the objective altogether or is blocked by a variable diaphragm inside the objective. There is thus no zero-order contribution to the image and the background is dark: only the weak diffracted beams are collected. It is therefore desirable to eliminate all stray light, and the alignment procedure described in Chapter 1 is particularly important. A dark-field condenser for high-power work may require to be 'oiled' to the bottom of the slide, and the adjustment is rather critical; it is best to start with the iris diaphragm (Fig. 27) opened up to admit light, until the preliminary alignment is complete.

By making the illumination even more oblique than in Fig. 27, one can arrange that it undergoes *total internal reflection* at the interface between a glass slide and the liquid medium containing a specimen. Under these conditions, scattered light arises only at points of close contact between the slide and any objects contained in the medium. This technique is described in more detail at the end of Chapter 8, where other methods of visualizing cell adhesion sites are also considered.

Questions

1. Since transparent objects can be made visible by stopping down the illumination, why is this simple method not fully satisfactory?

2. In what circumstances might a transparent object remain invisible in a phase microscope?
3. In the case above, how could the object be made to appear visible?
4. How could the refractive index of a small object be measured with the aid of a phase microscope?
5. How can the scattering of light by a very small object be explained by wave theory?

6 Fluorescence microscopy

The growing importance in biology of fluorescence microscopy is due to the remarkable selectivity that is now possible; one can link a fluorescent dye to a specific antibody, raised against the particular macromolecule of interest. The technique is also far more sensitive than conventional staining methods, and with its use some hitherto unsuspected details of cellular architecture have been revealed. Because of the low intensity of most fluorescence, the optical system usually incorporates dark-field illumination (Chapter 5) with some additional components to absorb or reflect certain wavelengths. To understand why these are necessary we must first go into the photochemistry of the method.

The nature of fluorescence

It has been known for a long time that certain substances will absorb a quantum of light of one wavelength, and after a certain delay re-emit a quantum which is usually of a longer wavelength. When the delay between absorption and emission is less than one-millionth of a second, the effect is called *fluorescence*; when the delay is much greater than this it is called *phosphorescence*. Both are examples of *luminescence*, which includes all cases where at least some of the absorbed energy is given out as light instead of being dissipated as heat. There is in general an increase in wavelength on re-emission because the energy of a quantum of radiation is proportional to its frequency ($E = h\nu$, where h is Planck's constant); since some energy is always lost, the frequency of light emitted must be lower and the wavelength therefore longer (Chapter 2). Exceptions arise only when collisions between molecules impart extra energy to the emitter.

Many natural materials fluoresce when illuminated with ultraviolet light, but this is of limited value in cell biology because it is impossible to distinguish many components from each other; for instance, the fluorescent amino acid tryptophan occurs in the majority of proteins. This effect is called *autofluorescence*, to distinguish it from the use of fluorescent dyes or *fluorochromes*. These may be linked to specific antibodies as described later. Of the dyes most commonly used at present, fluorescein isothiocyanate (FITC) absorbs blue light and emits green, while rhodamine compounds absorb green light and emit red. In both cases there is a broad range of wavelengths capable of exciting fluorescence, and another broad band of emitted wavelengths; the *absorption and emission spectra* define these ranges for a given substance.

The efficiency of energy conversion varies within the absorption spectrum, but for maximum total effect one clearly needs to excite over as broad a band as possible. One similarly needs to collect all the emitted wavelengths. At the same time one cannot allow the exciting radiation to extend into the range of wavelengths collected, because the incident light may be scattered by the specimen to generate an unwanted background. There is therefore a need for wavelength filters having sharp cutoffs and low absorption within selected regions. Recent technical advances in this field have greatly increased the sensitivity of the method.

Types of wavelength filter

Early types of filter (still available because they are cheaper) were simple absorbers of light, made of coloured glass. However, in order to cut out both ultraviolet and long-wavelength components from the exciting radiation, one needs to use a *band-pass filter*. This is made by combining a filter passing long wavelengths with one passing short wavelengths, giving a pass-band where their transmission curves overlap. With the broad cutoffs characteristic of absorbing filters this pass-band has poor discrimination and low transmission, even at the centre of the band.

Interference filters help to overcome these problems; as Fig. 28 shows, one can then achieve very sharp cutoffs with high transmission in the selected region. The best type of interference filter consists of a sheet of glass upon which are deposited several very thin layers of transparent

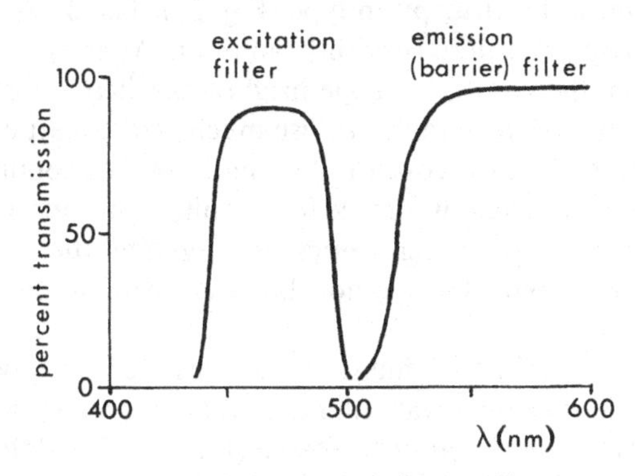

Fig. 28. Transmission by interference filters.

materials having different refractive indices. The mathematical explanation of their action is lengthy, but the effect is to produce partial reflection of light at each boundary between layers; all the reflected (or transmitted) beams then combine to interfere constructively for some wavelengths and destructively for others (Fig. 7, p. 13). Iridescent scales using this principle are common in nature. Another type of filter consists of one transparent layer between half-silvered surfaces, but this suffers from low transmission. Thus, interference filters can be used for reflection as well as for transmission.

The commonest (and simplest) man-made example of the transparent-layer type is the anti-reflection coating or 'blooming' which is now used for all but the cheapest cameras, binoculars and microscope objectives. It consists of a single layer, of optical path length about one-quarter of a wavelength, which reduces reflection over a broad range of colours; only the blue and red extremes of the spectrum remain, to give the characteristic purple colour seen by reflected light (one can always identify an interference filter by its different appearances in transmitted and reflected light). Another common application is the transparent *heat filter* used with mercury lamps; it does not absorb heat, but reflects both infrared and ultraviolet radiation back towards the source. For this reason, such a filter has to be placed with its interference layers facing the lamp; there will otherwise be excessive absorption of heat by the glass backing. Incidentally, all mercury lamps should incorporate an ultraviolet filter to avoid possible damage to the eyes.

Optical systems for observing fluorescence

The earliest method (still used for some applications) employs transmitted light, with dark-field illumination (Chapter 5) at high magnifications to reduce unwanted background. The system employs an *excitation filter* and a *barrier filter* of the absorption type (Fig. 29). The barrier filter cuts out most of the direct or scattered exciting radiation. Apart from losses due to filter absorption, there is poor image brightness when a dark-field condenser is used because of its inability to use much of the light coming from the source (due to the hollow cone of illumination). One commercial model has incorporated an ingenious device for simultaneous observation by phase contrast, using polarizing optics to regulate the level of background illumination, but this cannot be used with a dark-field illuminator.

With the advent of interference filters a new arrangement became practicable in which the specimen was illuminated from above; this was called *incident-light excitation* or *epi-fluorescence* (Fig. 29). In addition to the two filters there is a *chromatic beam splitter* (sometimes referred to

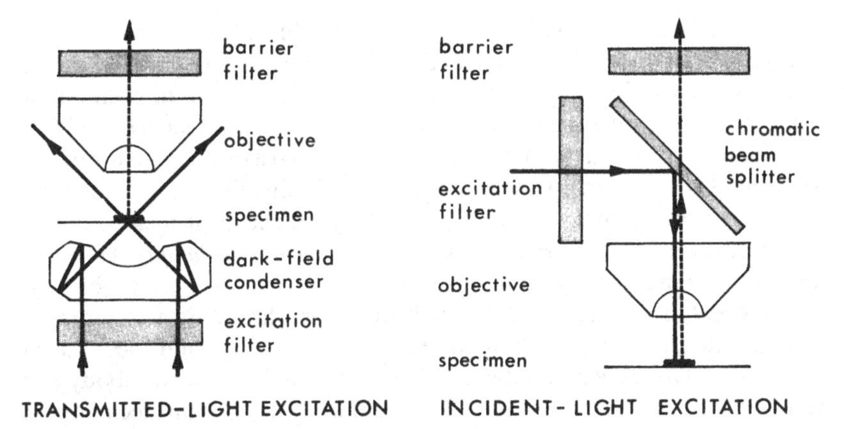

Fig. 29. Optical arrangements for fluorescence microscopy.

rather misleadingly as a 'dichroic mirror'). This is an interference filter inclined at 45° to the incident light, and having the property of reflecting short-wavelength radiation and transmitting longer wavelengths; its characteristics are similar to those of the barrier filter in Fig. 28. It reflects both the exciting radiation and any back-scatter from the specimen, but transmits to the observer all the fluorescent radiation above the cutoff wavelength. This extra selectivity, combined with other factors already mentioned, gives a considerable increase in sensitivity; it is also possible to illuminate the specimen normally from below while setting up, using phase-contrast optics.

Some further technical matters are worth noting. The objectives used in this work need to have the highest possible numerical aperture so as to collect the faintest fluorescence; they also have to be made of glasses that do not themselves fluoresce. The microscope lamp obviously needs to emit strongly in the excitation band of the fluorochrome, and for some dyes such as FITC a mercury lamp is not very good; a bright tungsten halogen lamp is often used instead. *Fading* of fluorescence is an irreversible photochemical effect caused by the exciting radiation, and one needs to limit exposure of the specimen by means of a shutter. *Quenching* is a reduction of fluorescence due to the presence in the specimen of interfering substances, such as certain metal ions. In this connection the pH of the specimen may also be important.

Immunofluorescence

Although this topic does not involve any optical principles other than those mentioned, the chapter would be incomplete without a brief description of such a powerful technique. In the clinical field it enables virus antigens to be located, so that diagnosis of infection is possible at the subcellular level. In cell biology it enables the distribution of even minor components to be mapped. The principle is that a marker such as FITC is chemically linked to an antibody, which is then used as a 'stain' to locate the antigen. The antibody can either be raised directly against material extracted from similar cells and purified by biochemical means, or the sensitivity can be increased by use of the *sandwich technique*. Here the antibody is raised by injection of purified antigen into (say) a rabbit, but this antibody is not labelled. Instead, the fluorochrome is attached to a non-specific anti-rabbit-globulin antibody, raised in a second species such as the goat. The specimen is then washed first with the rabbit antibody and secondly with that from the goat. The main advantage of this technique is that each rabbit antibody molecule can bind several molecules of goat antibody, thus increasing the amount of dye bound. An added convenience is that the non-specific labelled antibody can be produced commercially in bulk, and stored until needed by the user. Fig. 30 illustrates some of the results obtained.

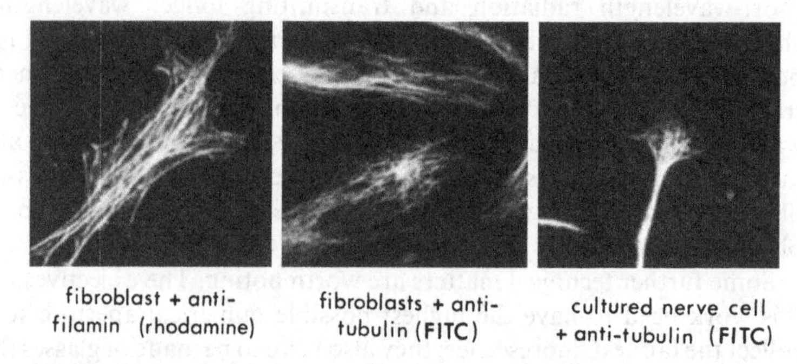

Fig. 30. Examples of immunofluorescence with the sandwich technique.

Other techniques

The immunofluorescence techniques described above can be used to *double label* a specimen, by linking one antibody to (say) FITC and a different one to rhodamine. By changing the whole set of wavelength filters one can then compare the distributions of two components.

An older and simpler technique achieves the same end in a limited number of cases. The dye acridine orange binds to all nucleic acids, but its fluorescence depends on the secondary structure of the substrate. Non-helical RNA structure encourages 'stacking' of the dye molecules, and this shifts the colour of the fluorescence from its normal green colour to a quite distinct red. Dye bound to double-helical DNA gives a green fluorescence. The quantity of dye needed is so small that cells can remain alive after its application.

A recent development called *total internal reflection fluorescence* is used to study the contacts between a cell and its substrate. Because the optical arrangement is similar to that used for interference reflection microscopy, details are given at the end of Chapter 8.

Questions

1. How can a mercury discharge lamp be said to be unsuitable for exciting FITC fluorescence, when its output is so much higher than that of a tungsten microscope lamp?
2. In the use of a chromatic beam splitter (Fig. 29), what difference would it make if the beam splitter was turned over to interchange its two faces?
3. If an interference filter is held up to the light and tilted, the transmitted colour changes slightly. Why is this?
4. You are trying to locate a minor protein component in a cell. Under which circumstances might the results obtained with immunofluorescence be misleading?

7 Polarizing microscopy

The usefulness of the polarizing microscope is not limited to the study of crystalline materials, though it certainly finds many applications in that field. There are many textbooks on crystallographic uses, and we shall not attempt to cover the same ground. This chapter is addressed to the microscopist who is mainly interested in determining molecular orientation, whether in biological or synthetic materials. The polarizing microscope can in this field be superior to any electron microscope. Typical applications to biology have included the study of cell spindle formation, and the behaviour of the components of a contracting muscle. Polarization measurements are also made in conjunction with fluorescence microscopy (Chapter 6), when information can be obtained about the orientation of a marker dye. In addition to these direct applications, polarizing optics form the basis of two common types of interference microscope (Chapter 8), and the contents of this chapter are essential to an understanding of how they work.

The production of polarized light

At the beginning of Chapter 2 (see Fig. 3) we noted that the electric field of a light wave oscillates at right angles to the direction of propagation, and that for plane-polarized light it is confined to one plane. Later we discussed the interference of two waves on the assumption that they had the same polarization. If, however, two waves are polarized at right angles to each other they are not able to interfere – just as the two tracks of a stereo gramophone record are independent, because one consists only of vertical displacements in the groove and the other only of lateral displacements. In the same way, a pair of light waves polarized at right angles to each other can travel the same path while each carries independent information.

In light from most sources there is a mixture of waves having all possible polarizations, and the first step in polarizing microscopy is to filter the light so that there is only one *vibration axis* for all light entering the microscope.

This is achieved nowadays by using a sheet of a material such as Polaroid. This is a plastic composed of oriented macromolecules, or in the early type tiny crystals embedded in plastic, which have the property (called *dichroism*) of absorbing light very strongly except for one direction of polarization. A sheet of Polaroid looks grey when held up to the light because about half the intensity is lost in this way.

Fig. 31 illustrates the principle; in this chapter it is important to note that we are now using a different application of the vector concept, in which the direction of the arrow represents the *direction in space* of the electric-field vibrations, instead of the phase angle as before. The length still represents the magnitude. In the present situation we can say that any vector is equivalent to a pair of new vectors in space, provided that they make up a parallelogram as before (Chapter 2). In particular, we can take any pair of

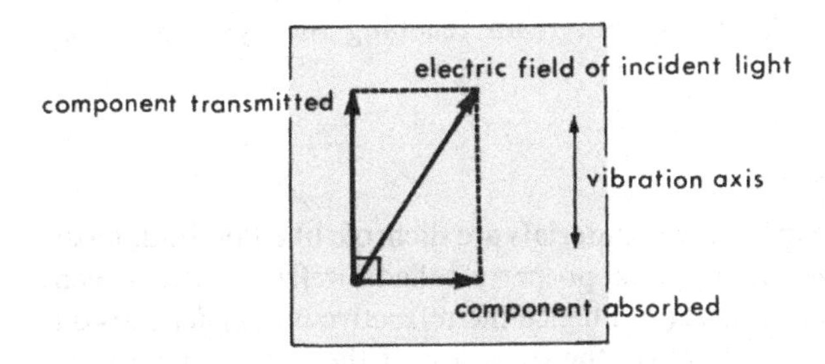

Fig. 31. Resolution into components at a dichroic sheet.

lines making an angle of 90° with each other, and by dropping perpendiculars onto these lines we can generate an equivalent pair of vectors called *components*; the original vector has been *resolved* along the chosen pair of directions. In Fig. 31 the chosen field directions are parallel and perpendicular to the vibration axis of the Polaroid; the light is assumed to be travelling at right angles to the surface of the sheet. Because of the dichroism of Polaroid, only one component gets through and the other is absorbed.

Fig. 32 shows how dichroic sheets are used in a polarizing microscope. The first sheet is called the *polarizer*; it is generally placed underneath the condenser, and is either fixed in orientation or rotatable between four 'click' positions spaced 90° apart. At each of these positions the vibration axis is either 'north–south' or 'east–west' with respect to the microscope stage as viewed by the operator. The dotted vector in the figure indicates that polarization can be either 'up' or 'down'; generally only one arrow is drawn, since the electric vector of any wave goes through both positive and negative half-cycles.

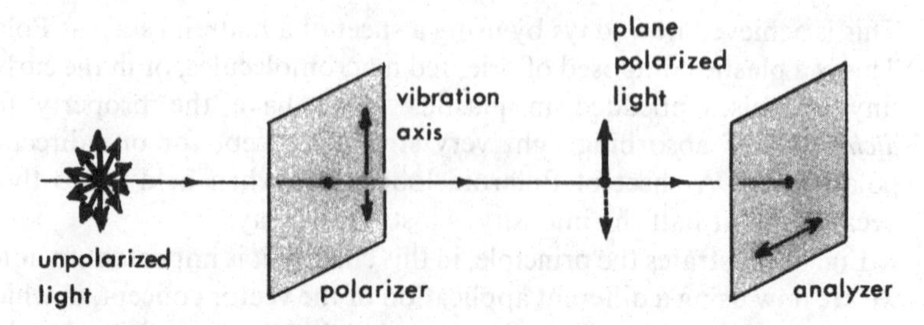

Fig. 32. Polarizing components of a microscope.

The second dichroic sheet in Fig. 32 is called the *analyzer* and it is generally arranged as shown with its vibration axis *crossed* with that of the polarizer. In a microscope it is placed below the eyepiece, and has the effect of preventing all background light from reaching the eye, just as in dark-field illumination.

Birefringence

Although some natural materials are dichroic like Polaroid, most are not; many, however, show the property called birefringence. In such materials the velocity of light (and hence the refractive index) depends on the plane of polarization, and on the direction of the ray relative to the specimen. In true crystals the situation can be very complex, but we are concerned here with simple cases such as the preferential orientation of long molecules along one direction. This can be induced in a stretched fibre or film, or may occur naturally in a biological organelle. In such a case there is a non-random orientation of chemical bonds, and the electric field interacts more strongly with the molecules when it vibrates along a preferred direction. This effect is called *intrinsic birefringence*. There is another kind, often combined with the first, called *form birefringence*; this can arise, for instance, when the structure consists of parallel rods immersed in a less-refractile medium (as in a striated muscle). The two kinds of birefringence can be separated by varying the refractive index of the medium; in the case where both contributions have the same sign (see later), the total birefringence passes through a minimum at a point where the form birefringence is reduced to zero.

In Fig. 33, light polarized parallel to the long axis of the fibre travels with one velocity, $v_{\parallel}$, while light polarized parallel to the short axis travels with a different velocity, $v_{\perp}$. (For light travelling down the axis of the fibre there is no such variation, unless the molecules form a three-dimensional crystal.) We have illustrated the situation where $v_{\parallel} < v_{\perp}$; since refractive index is inversely proportional to velocity (Chapter 2) it follows that $n_{\parallel} > n_{\perp}$. The

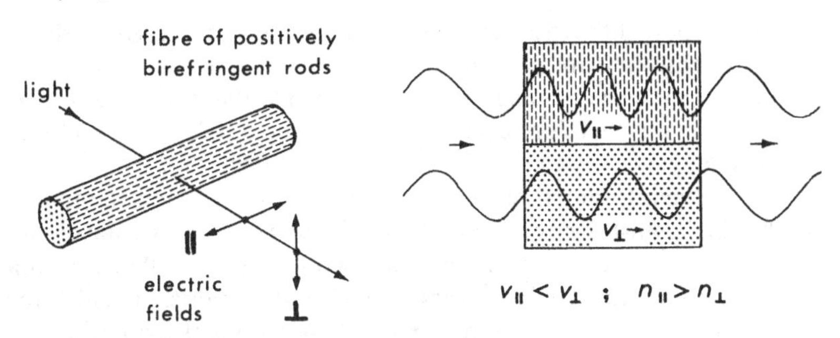

Fig. 33. Positive birefringence in a fibre.

definition of birefringence is b.r. $=(n_{\parallel}-n_{\perp})$, so in the case shown the birefringence is *positive*. In this case the long axis is known as the *slow axis* and the short one is the *fast axis*; for negative birefringence the opposite is true.

Fig. 33 also shows that, because of the difference in velocity, two waves polarized parallel to the fast and slow axes, and initially in phase with each other, will emerge from the fibre with an optical path difference (or phase difference) between them. We shall see that such an effect can make the object visible against the dark background of a polarizing microscope.

In Fig. 34 the same fibre has been placed with its long axis at 45° to the vibration axes of the crossed polarizer and analyzer. The light incident on the fibre must now be considered as broken up into two equal wave components, polarized along its fast and slow axes. Since these travel with different velocities they emerge with an optical path difference between them, denoted by Δ in the diagram. When the two waves strike the analyzer, it passes only the component of each wave along the analyzer's vibration axis; the other components (vertical in this diagram) are absorbed. The components passed by the analyzer, being similarly polarized, can now interfere with each other. In the absence of an

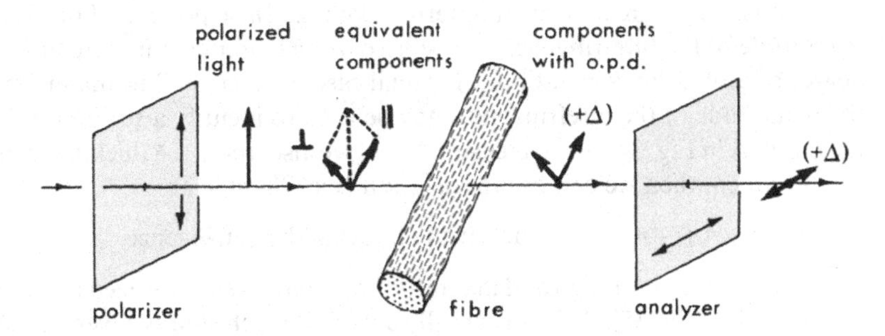

Fig. 34. A birefringent object in the polarizing microscope.

optical path difference between them they would cancel each other out, but in general they are not equal and opposite – one, for instance, may at a given instant be passing through zero while the other is at a maximum. They do not therefore cancel out, *unless* the o.p.d. is an exact number of wavelengths (we shall return to this point later).

In general, therefore, some light will be passed and the fibre appears bright against a dark background. If, however, it is rotated through a complete circle by turning the microscope stage, it will pass through four positions in which it disappears. In each of these positions, either the fast or the slow axis (it does not matter which) lies along the vibration axis of the polarizer, and one of the component vectors has zero amplitude; in Fig. 34, for example, the 'perpendicular' component will be zero when the fibre axis is vertical in the diagram. The remaining 'parallel' component is then blocked by the analyzer. A specimen is normally set up by finding one of these positions of *extinction*, and then turning the stage through exactly 45° to obtain maximum brightness.

The measurement of birefringence

The mere existence of birefringence does not tell us which way the molecules are oriented; in Fig. 34 the fibre would appear equally bright if the stage was rotated by 90°. To resolve this ambiguity we need to measure the *sign* of birefringence. If we already know whether the molecules making up an object are positively or negatively birefringent, then measurements on the specimen to determine which of its axes is 'slow' will tell us how the molecules are arranged; conversely, we could determine the sign of the molecular birefringence if (as in a fibre) it was reasonable to assume that the long axes of the molecules lay parallel to the long axis of the specimen.

It may also be useful to determine the *magnitude* of the birefringence, since this gives information on the amount of orientation within the specimen. The birefringence of chromosomes, for instance, is so weak that this rules out any simple model in which DNA runs parallel to their length; there must be a great deal of internal coiling. In a polymer fibre the magnitude of the birefringence can be used as an empirical measure of the degree of molecular orientation. In special cases (e.g. crystalline materials) the magnitude of the birefringence may be used to identify a substance. If the o.p.d. Δ in Fig. 34 can be measured, and we also know the thickness t of the specimen, then we can use the relation (see Chapter 2).

$$\Delta = t(n_{\parallel} - n_{\perp}) = \text{product of thickness and birefringence}$$

For measuring both sign and magnitude of birefringence we need to use a *compensator* (Fig. 35). This is generally a device which simply reverses the phase shift introduced by the object, so that it appears dark like the

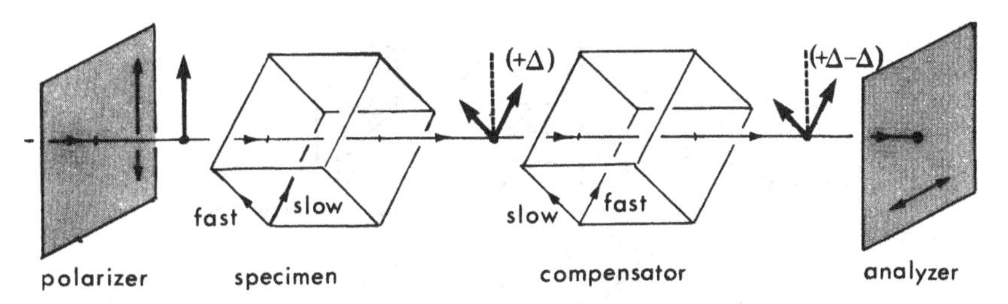

Fig. 35. The use of a compensator.

background. A compensator consists essentially of a birefringent crystal plate, arranged (like the specimen) with its fast and slow axes at 45° to the vibration axis of the polarizer. In a polarizing microscope there is a slot for the compensator somewhere between polarizer and analyzer, the direction of the slot being preset at 45° to the 'north–south' direction of the stage. If the slow axis of the compensator (sometimes marked on it by a line labelled 'γ') lies parallel to the *fast* axis of the specimen, then it is in a position to reverse the phase difference.

If the o.p.d. introduced by the compensator is exactly equal and opposite to that arising from the specimen, the components passed by the analyzer will then cancel each other so that the object appears black. Most compensators are made so that the o.p.d. can be varied, and they are calibrated accordingly. They are operated with white light (for reasons discussed below) and the setting is changed until the object goes black. If this proves impossible, it may be that the slow axis of the compensator is parallel to the slow axis of the object, so that the o.p.d. is only made greater. In this case one needs to rotate the specimen stage by 90°, thus exchanging fast and slow axes, before trying again.

There are many designs of compensator, covering different ranges of o.p.d. from a fraction of one wavelength to a hundred wavelengths or more. The theory behind some of them is rather complex, but there are two simple kinds that we will discuss here. The first is the *quartz wedge* (Fig. 36). This is a plate of variable thickness which is moved in and out of the compensator slot until an appropriate region intercepts the light coming from the specimen.

Calibration is performed by changing to monochromatic light and noting (in the absence of a specimen) the positions of black bands or *fringes*, each of which denotes the existence of a whole number of wavelengths of o.p.d.; $\Delta = N\lambda$ where N is a whole number. This gives extinction (blackness) because the two components passed by the analyzer are equal and opposite. The positions of the calibration bands vary with the wavelength. In white light the fringes are therefore not black but

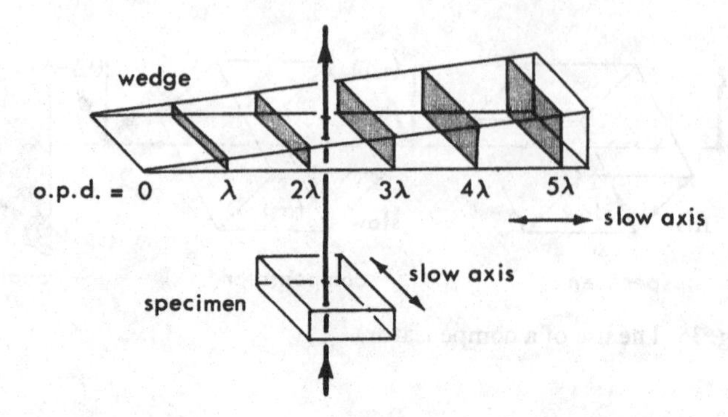

Fig. 36. A quartz wedge compensator in monochromatic light.

multicoloured, *except* at zero path difference ($N=0$). This explains why compensation has to be done with white light; in white light there is only one situation (zero net path difference) in which the object goes black, whereas in monochromatic light the object will go black at intervals of one wavelength o.p.d. along the wedge. Only the fractional part of the o.p.d. can then be determined. If in the example illustrated by Fig. 36 the specimen goes black in white light at the position shown, its o.p.d. must be about 2.5 wavelengths of the light used for calibration.

Another kind of compensator called the *first-order red plate* is useful when the specimen o.p.d. is a small fraction of a wavelength. The plate is not variable in o.p.d., but relies on colour changes (to which the eye is very sensitive) to determine the sign and magnitude of birefringence. It consists of a uniform crystal plate whose o.p.d. is equivalent to exactly one wavelength of light in the green/yellow centre of the spectrum, so that in such light it gives a dark field. In white light, however, the red and blue ends of the spectrum get through the analyzer to produce the purple-red colour that gives the plate its name.

If now a suitable specimen is placed in the field with its slow axis parallel to that of the plate, the colour shown *by the specimen* will be shifted towards a blue-green colour, which comes from subtracting a different band from the spectrum of white light (Fig. 37). If the specimen stage is turned through 90°, the colour changes to an orange hue. By noting the orientation of the long axis of the specimen (if it has one) relative to the slow axis of the plate (marked on it) one can then determine the sign of the birefringence. An approximate value of the magnitude can also be obtained if needed, by matching the specimen colour to a suitable colour chart (available from microscope manufacturers).

The words 'first order' in the description indicate that the plate thickness corresponds to the first band in the series shown (for a wedge) in Fig. 36. If the o.p.d. of a specimen is greater than about one-third of a wavelength,

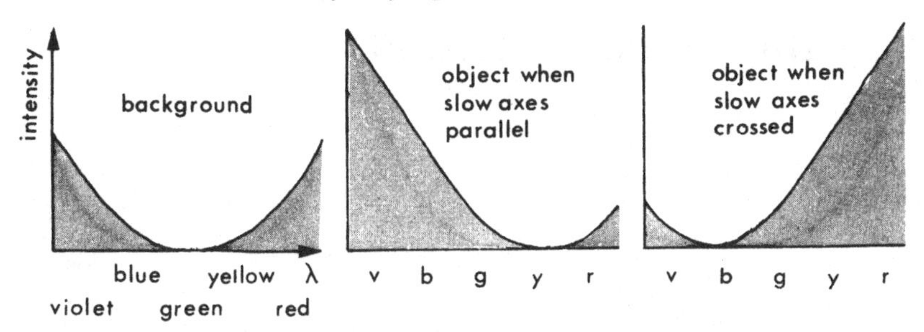

Fig. 37. Colours produced by the first-order red plate.

there is a danger that the colour will be that corresponding to a higher-order set of fringes: instead (say) of moving into the first-order blue region, one may shift to the second-order orange. This would give wrong answers for both the sign and the magnitude of the birefringence. It is therefore *essential* to check first with a wide-range compensator such as a quartz wedge, to make sure that the o.p.d. is suitably small.

A promising recent development has been the use of video techniques to study objects of very weak birefringence. By using a video camera with an appropriate display system one can greatly enhance the contrast of the image, since it is possible to electronically subtract the background level of illumination. By amplifying the signal before display, one can also improve the sensitivity and reduce recording times. The same technique can usefully be applied to interference contrast microscopy (Chapter 8).

Questions

1. If between crossed polarizer and analyzer (Fig. 32) we place a third dichroic sheet with its vibration axis at 45° to that of the polarizer, will the analyzer pass any light?
2. If a permeable specimen has a large, positive intrinsic birefringence and a weaker, negative form birefringence, how will the total birefringence change as the refractive index of the medium is changed?
3. If a fibre consists of negatively-birefringent molecules, lying parallel to its length, which is the slow axis?
4. If a fibre of thickness 0.1 mm extinguishes when the compensator o.p.d. is one wavelength of green light (546 nm), what is the magnitude of its birefringence?
5. Why does the monochromatic fringe spacing in a quartz wedge vary with the wavelength of the light?

8 Interference methods

The commonest modern application of interference techniques to the study of transparent objects is known as *interference contrast*; this is an alternative to phase contrast, and is preferred by many users because there is no 'halo effect'. An older use of the same principle is in the Jamin–Lebedeff interference or *interferometer* microscope, which is used for making quantitative measurements of mass or thickness. Both types use polarizing optics as a means to an end, and we shall assume that the contents of Chapter 7 are understood. There are other kinds of interferometer microscope based on different optical systems, but most of them are now largely obsolete and they will not be considered here; they are described in detail in some of the older text-books. We shall, however, consider the new technique of *interference reflection*.

Interference contrast

There are several variants which work on the same general principle. A commonly-used system is the adaptation, due to Nomarski, of the interferometer microscope described later. It is not suitable for quantitative measurements, but for thicker objects it produces a 'cleaner' image than the phase microscope; one can also obtain striking colour contrast effects. It can be used with either transmitted or reflected light, but we shall discuss only the former. It is relatively expensive, and for very thin specimens the contrast may be inferior to that seen with a phase microscope.

In the optical system (Fig. 38) the first step is to produce plane-polarized light, which then strikes a *beam splitter* combined with the condenser. The beam splitter is a double prism made of a crystalline material showing *double refraction*; incident light is split into two beams, following different paths and polarized along directions at right angles to one another. By arranging these two directions to make angles of 45° with the vibration axis of the polarizer, we produce two coherent components forming broad beams of light with a lateral displacement or *shear* between them. The shear

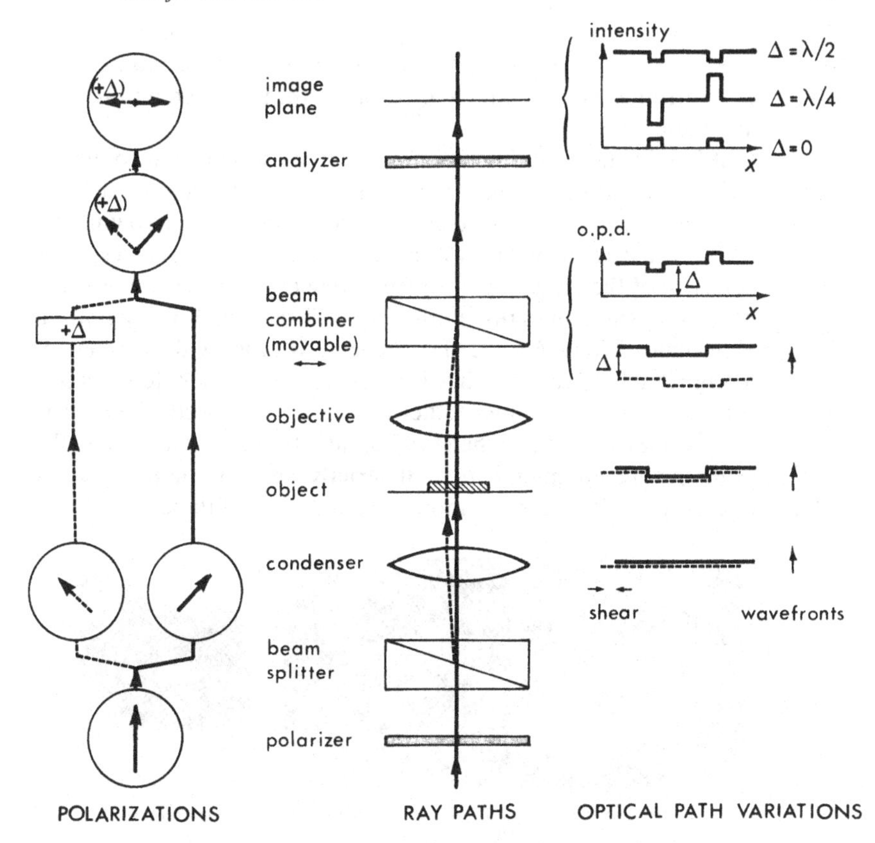

Fig. 38. The optical system for interference contrast.

has been greatly exaggerated in Fig. 38; in practice it is comparable to the resolution limit of the objective, and a different double prism is used for each objective.

Placed in a slot above the objective (or sometimes combined with the analyzer) is a second double prism, the *beam combiner*. This not only recombines the two beams, but in some instruments it also introduces a constant o.p.d., Δ, between them, which can be varied by sliding the beam combiner in and out of its slot. One effect of this is to change the amount of background light passed by the analyzer. This can be seen from the left-hand diagram in Fig. 38, which shows the relative polarizations when viewed down the beam; as in Chapter 7, we are using the direction of a vector to represent the plane of polarization. When $\Delta = 0$ the background will be dark, while $\Delta = \lambda/2$ (equivalent to *reversal* of the dotted analyzer component) gives maximum brightness for light of wavelength λ. In white light the background then shows a strong colour, which can be changed by small movements of the beam combiner. Variations in colour can also be obtained by introducing a 'λ plate' or by rotating the polarizer, but we shall not discuss these here.

The right-hand diagram in Fig. 38 shows that the specimen introduces a 'trough' of optical path difference into both beams, but when the shear is removed these troughs no longer exactly overlap. For the idealized uniform specimen illustrated, there will now be an extra 'blip' of o.p.d. corresponding to each *edge* of the specimen, positive at one edge and negative at the other. When $\Delta = 0$ this means that both edges will show up bright against a dark background. When $\Delta = \lambda/2$ the converse is true, with dark edges against a bright background – or, in white light, edges of one colour against a different coloured background. Other effects are produced when Δ has intermediate values; one edge is then brighter and the other darker than the background, giving an interesting but strictly spurious impression of 'shadowing'. Fig. 39 illustrates (in monochrome) some of these effects, and

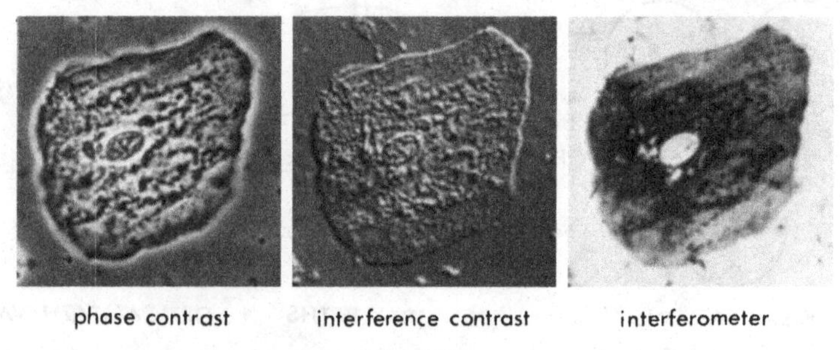

Fig. 39. A human epithelial cell in different types of microscope.

compares the images of the same object in the phase microscope and in the interferometer microscope described below. The reason for making the shear comparable to the resolving limit is now apparent: it shows up the edges without producing the diffuse halo characteristic of phase microscopy.

The interferometer microscope

This type is often referred to simply as the 'interference microscope', but there is an important difference from the Nomarski system just described: the two beams are separated by a distance that is a sizeable fraction of the field diameter. Fig. 40 shows the optical system. The birefringent beam splitter produces an 'ordinary' beam which passes through it undeviated, and an 'extraordinary' one which disobeys the normal law of refraction and goes off at an angle to the first beam. The two

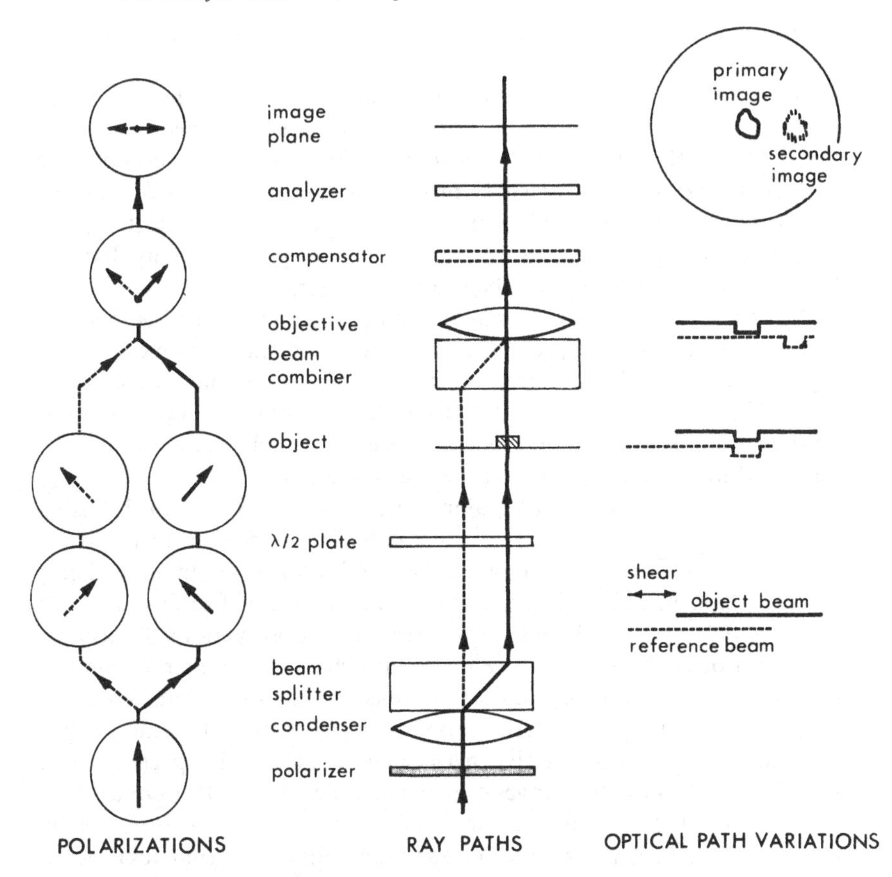

Fig. 40. The Jamin–Lebedeff interferometer microscope.

beams are polarized in different directions as shown. In order to equalize the optical paths of the two beams it is necessary for them to exchange polarizations, so that the 'ordinary' from the beam splitter becomes the 'extraordinary' in the beam combiner. This is achieved with the aid of a *half-wave plate* which effectively rotates the planes of polarization in opposite directions by 90° each.* As the name of the plate implies, it is designed for a specific wavelength such as 546 nm; for reasons we shall not

* For those who are interested, it works as follows: the slow axis of the plate makes an angle of 45° with the directions of polarization of the two beams, and each is resolved into a fast and a slow component. A half-wave path difference is introduced by the plate between these two components, and this is equivalent to reversing the direction of polarization of one component. Both resultant vectors are thus rotated by 90°.

go into, additional 'ghost' images may be produced when white light is used.

When the shear between the two beams is removed by the beam combiner, the 'troughs' of o.p.d. produced by a small object no longer overlap at all. Each will therefore give rise to an image; however, only one of these (the *primary image*) will be properly focused. The *secondary image*, corresponding to a 'trough' in the reference beam (shown dotted), will show serious astigmatic distortion, because the image-forming rays have all been bent in one direction by the beam combiner.

We can now see another advantage of using polarizing optics: one can use a compensator (Chapter 7) to measure the o.p.d. introduced by an object, and hence deduce mass or thickness from the magnitude of the o.p.d. There is of course one limitation, that birefringent objects such as crystals can give misleading results; one needs to check whether or not the object is birefringent by rotating the specimen stage.

There are various ways of making measurements. One can, by using controls that tilt and rotate the beam splitter and combiner, respectively, produce a uniform background which is either dark or (in white light) shows a uniform colour. In the latter case an object will show a different colour which gives good contrast for qualitative observations. By using a compensator such as the quartz wedge described in Chapter 7, one can make first the background and then the object go black (in white light), and thereby estimate the o.p.d. between object and background. In monochromatic light the quartz wedge gives a *banded field* of regularly-spaced fringes, and one can then use the fringes crossing an object to produce contours of equal o.p.d. (Fig. 41). In the same way, thick objects (such as the oil droplets illustrated) will show contours even with a uniform background. Each contour represents a wavelength increment in o.p.d. To obtain an accurate value of the fractional-wavelength part of an o.p.d., one can use a *Sénarmont compensator* in which a *quarter-wave plate* is placed in the

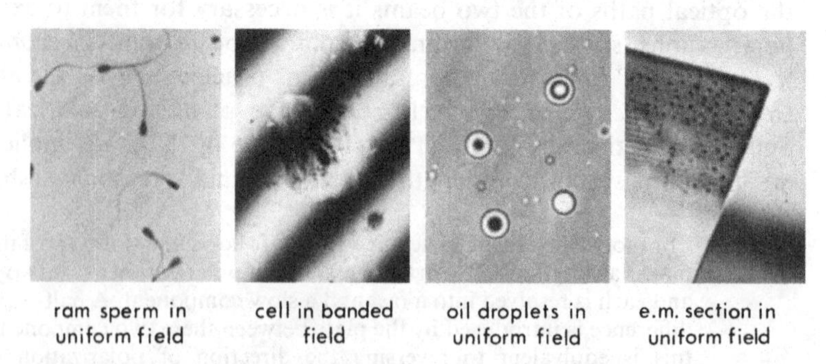

Fig. 41. Objects seen with the interferometer microscope.

compensator slot, and compensation is achieved by rotating the analyzer. The theory behind this is too complex to be dealt with here.

The extreme right-hand picture in Fig. 41 illustrates a problem that can arise: we cannot make quantitative measurements if the primary image is overlapped by the secondary image either of another object, or of a different region of the first one. The method is therefore unsuitable in certain cases; one must have a small object well separated from others, or one in which only the region near the edge needs to be studied.

The measurement of mass or thickness

The interferometer microscope described above enables us to measure the o.p.d. introduced by an object relative to the medium in which it is immersed; only the *dry mass* is relevant, and the o.p.d. gives the mass *per unit area* of the region studied. One therefore needs also to measure the area to obtain the total mass, and if the o.p.d. is non-uniform some kind of integration must be performed. There are specialized flying spot microscopes that enable this to be done automatically. We are concerned here only with the basic optical principles of the method.

To obtain the mass per unit area we do not need to know the thickness of the specimen; it does not matter whether the solid material is concentrated or dispersed – all that matters is the *total extra mass* through which the specimen beam passes. One needs also to know the *specific refractive increment* which is the increase in refractive index produced by a one per cent change in concentration of the specimen material.

The expression for mass per unit area can be derived as follows: consider a region of specimen having area A m^2 at right angles to the beam, and thickness t m (Fig. 42). Let the percentage concentration of dry mass in the

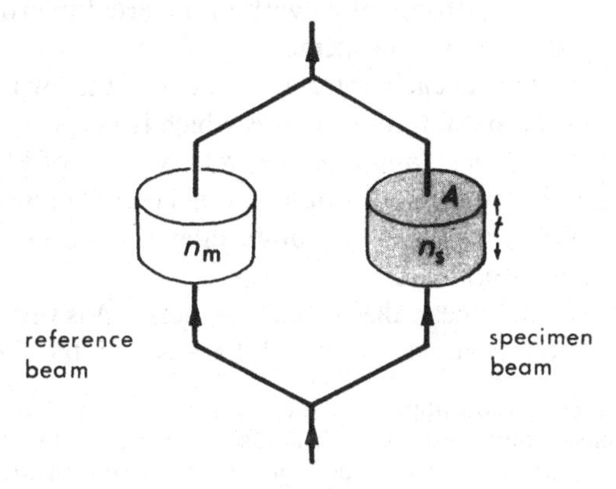

Fig. 42. Diagram illustrating the theory of quantitative measurements.

specimen be c; the units of this are grams per decilitre, so the concentration in grams per cubic metre is 10 000 c or $10^4\ c$. The total dry mass in the volume considered is then $M = 10^4\ cAt$ g, and the mass per unit area is $m = 10^4\ ct$ g m^{-2}. The o.p.d. is $\Delta = (n_s - n_m)t$ m, where n_s is the refractive index of the sample and n_m that of the medium in which it is immersed.

The specific refractive increment is defined as

$$\alpha = (n_s - n_m)/c = \Delta/ct$$

so that the mass per unit area becomes

$$m = 10^4\Delta/\alpha \text{ g m}^{-2}*$$

It is fortunate for biologists that α is almost the same (about 0.0019) for proteins and nucleic acids in water, and quite similar (0.0014) for sugars and lipids. The mass of an organelle containing a mixture of substances can therefore be determined without knowing its composition.

We can also determine the thickness of a thin film or sheet of solid material, if we know its refractive index; the method can be used, for example, to determine accurately the thickness of an electron microscope section. If the reference beam travels through air, the o.p.d. is $\Delta = (n-1)t$ and t can be determined.

Interference reflection microscopy

In this fairly new application of interference principles, the pattern of contacts between a cell and its substratum is made visible. Incident-light illumination is used as in fluorescence microscopy (Fig. 29, p. 43) except that the chromatic beam splitter is replaced by a simple half-silvered mirror, and monochromatic or white light is used throughout. The cells are generally made to stick to a glass coverslip which is then inverted for viewing from above, or viewed from below with an inverted microscope. Fig. 43 illustrates the first of these arrangements.

Light is partially reflected at each interface, and for the two waves illustrated there is an optical path difference of Δ which is proportional to the distance between cell and glass, and a further relative shift of $\frac{1}{2}\lambda$ which arises whenever light is reflected from the boundary of a region of increased refractive index. This applies to the cell–medium interface but not to the boundary between medium and glass.

At the adhesion sites (sometimes called 'focal contacts'), Δ is very small; the two waves will therefore tend to cancel because of the $\frac{1}{2}\lambda$ path

* A very similar expression applies to the measurement of stain uptake under bright-field conditions; in such a case m is proportional to the *absorbance*, defined as $\log(I_0/I)$ where I_0 and I are the incident and transmitted intensities.

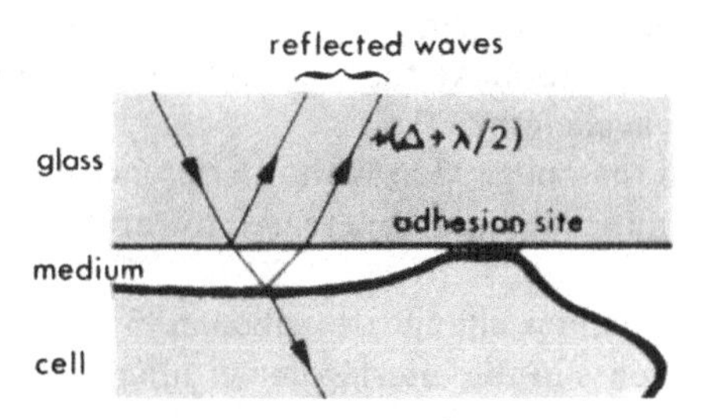

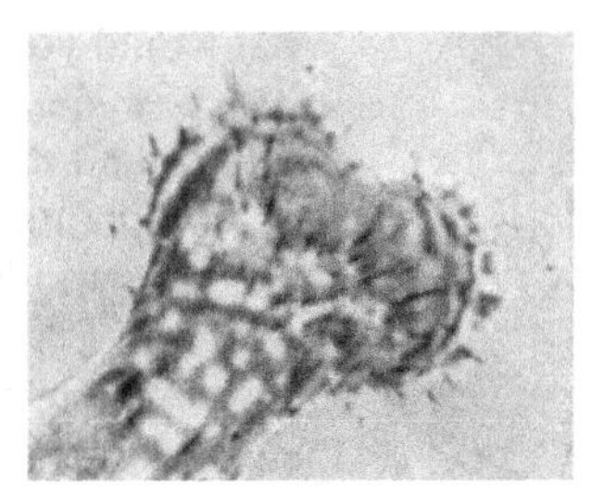

Fig. 43. Interference reflection microscopy.

difference, though their amplitudes are not equal and the resulting interference colour is grey rather than black. The adhesion sites will thus appear darker than surrounding regions. Fig. 43 shows the image given by part of a motile fibroblast; the adhesion sites show as fine dark lines near the edges of the cell.

Some optical systems employ polarized light and include extra components whose main function is to eliminate stray light reflected from lens surfaces; the basic principle is, however, the same. It is also sometimes necessary to distinguish the reflection effect shown in Fig. 43 from effects associated with the waves reflected from other regions of the cell (not illustrated). This can be done by varying the numerical aperture and wavelength of the illumination, and the refractive index of the medium.

Total internal reflection microscopy

Although nothing to do with interference, techniques in this class are best described here because the optical arrangement is so similar to that in Fig. 43. As in that case, the methods are used to study cell–substrate contacts. However, the angle of incidence is increased so that it is greater than the *critical angle* for a glass–water interface, at which no light is refracted down into the medium: it undergoes total internal reflection within the glass. Under these conditions the light only penetrates for a very small distance as an *evanescent wave*, but this is sufficient to excite scattering (or, in the most recent development, fluorescence from suitable marker molecules) at the cell–substrate adhesion sites. The thickness of the excited layer depends on the angle of incidence, and can be as small as 100 nm. If the arrangement is such that reflected light is not collected by the objective, adhesion sites then show up as bright regions on a dark background.

Questions

1. Why in the microscopes described is it necessary to have the two beams polarized at right angles to each other?

2. What would you expect to happen in the Nomarski system if the beam combiner is moved to make the constant o.p.d. Δ (Fig. 38) greater than half a wavelength?
3. What will happen to the image seen with interference contrast if the refractive index of the medium is increased towards that of the specimen?
4. How will the refractive index change described in (3) above affect the measurement of mass in the interferometer microscope?

9 Aberrations and microscope design

In all previous discussion of optical systems we have assumed that, apart from diffraction effects arising from limited apertures, all lenses form perfect images. For simplicity we have also represented an objective or an eyepiece by a single biconvex lens. The first microscopes were indeed like this, but they suffered severely from *aberrations* or geometrical distortions, including an inability to form a well-focused image in one plane. This arises because the lens formulae given in Chapter 1 apply strictly only to very thin lenses, and for rays travelling close to the axis. A modern high-power objective contains several relatively thick lenses and they are far from biconvex in shape, so as to reduce the aberrations; the front lens, for instance, often has a hemispherical form with its flat surface facing the specimen. In addition, glasses of more than one refractive index are used. This chapter begins with a description of the standard aberrations, so that the reader can recognize them in a badly-made microscope; we also describe a simple method of testing the resolution.

Chromatic aberration

This is the easiest aberration to understand. It arises from the fact that for any kind of glass the refractive index (r.i.) varies with wavelength (dispersion), so that the focal length of a single lens (determined by its r.i. and the curvature of the surfaces) cannot be the same for all colours. In white light, therefore, the image will show blurred and coloured edges due to each wavelength giving an image of a different magnification, in a different focal plane.

The effect can greatly be reduced by making an *achromatic doublet* (Fig. 44) in which a converging (convex) lens is cemented to a diverging (concave) lens made of a more dispersive glass; the diverging lens reduces the magnification, but the components are designed so that the doublet is still equivalent to a converging lens. In this way the focal length can be made the same for two colours; at other wavelengths, however, there are still small deviations. These can be somewhat reduced by combining a glass

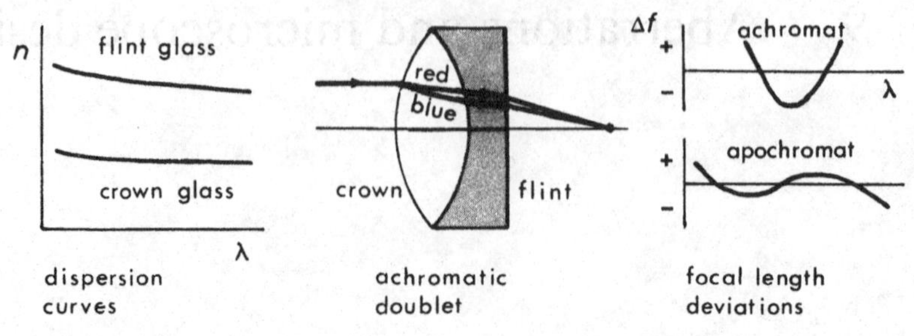

Fig. 44. Chromatic aberration.

and a fluorite lens, but the only way to obtain correction at more than two wavelengths is to use more than two kinds of glass. An *apochromatic* objective has the same focal length at three wavelengths (Fig. 44); it may still (like an achromat) fail to produce the same magnification at more than one wavelength, and one uses a *compensating eyepiece* for correcting this defect.

Spherical aberration

This is the most important of the *monochromatic aberrations*, which all arise from the fact that rays passing through different regions of a large-aperture lens do not come to the same focus. In a cheap camera these problems are avoided by reducing both the lens aperture (with a loss of camera speed) and the width of field; cheap binoculars also have a restricted field of view. Spherical aberration, which is a very serious problem in the design of electron microscopes, applies to objects on the axis; for example, rays travelling through the centre of a biconvex lens may focus some way beyond those passing near the perimeter (Fig. 45).* An improvement can in this case be achieved by substituting a meniscus or a planoconvex lens; the hemispherical front lens of many objectives is one variant of this.

One also has to allow for spherical aberration induced by the coverslip between specimen and objective. Fig. 45 shows that, from the point of view of the objective, the specimen *appears* nearer to it for extreme rays than for axial rays. If this is corrected for in the design of the objective, one must always use the right thickness of coverslip; quite serious spherical aberration will appear if the coverslip is omitted. Some objectives are made with a variable *correction collar* for different thicknesses of coverslip.

* Aberrations can be 'positive' or 'negative'; in the present case this means that extreme rays can focus either in front of, or beyond, the focal point for axial rays. Standard textbooks vary in which sign of aberration is illustrated.

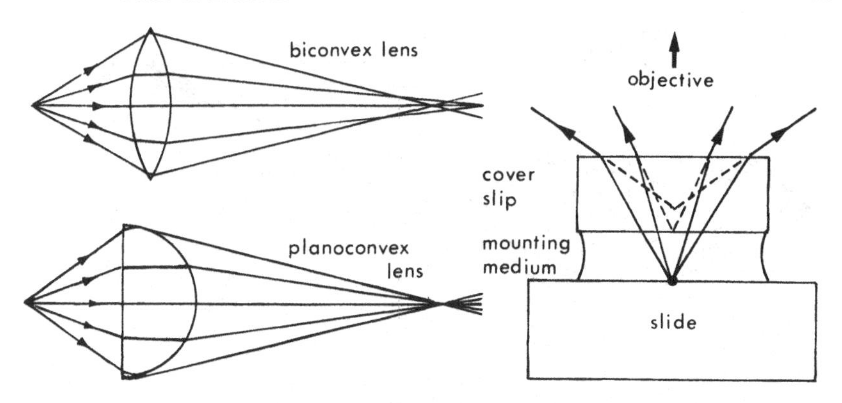

Fig. 45. Spherical aberration and its correction.

Other aberrations

All other aberrations arise when an object point lies off the axis of the lens. *Coma* is the failure to form a perfect image of a point near the axis (Fig. 46), for reasons similar to those causing spherical aberration. However, it is not as simple to correct; it may be necessary to make the lens surfaces slightly non-spherical (i.e. of curvature varying with distance from the axis). A good objective should show no trace of coma, but when present it causes a point to be imaged as an asymmetrical comet-like figure.

Astigmatism is also illustrated in Fig. 46. It arises even when there is a limited aperture, for object points *well off* the lens axis (it also crops up in electron microscopes, but for quite different reasons). Rays travelling in a plane containing the lens axis ('horizontal' in our perspective drawing) converge some way before those travelling in the 'vertical' plane. Instead of a point image there is a blurred 'circle of least confusion', on each side of which there is a sharp line, one horizontal and one vertical. The defect can be reduced by suitable spacing of lenses in an objective.

Field curvature is illustrated in Fig. 47, which shows drawings of the image of a rectangular mesh. It is possible to focus the outer region, but only by putting the centre out of focus. One can see this clearly when viewing a slide of bacteria with an ordinary × 100 achromat and a standard eyepiece. It is possible to compensate for the effect with a special eyepiece, or to design a flat-field objective; in this case, however, there will still be a need for an eyepiece that corrects for lateral chromatic aberration (variation of magnification with wavelength). Field curvature is clearly of most importance when the image is to be photographed.

Two other field distortions are illustrated in Fig. 47. The names are

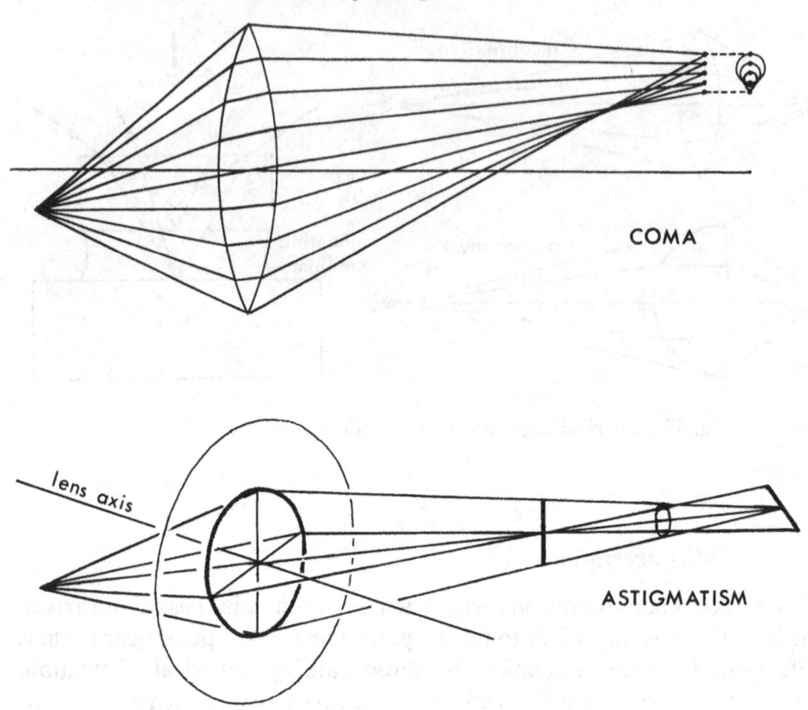

Fig. 46. Coma and astigmatism.

self-explanatory; the effects are important only when it is necessary to make accurate measurements of size or shape.

Objective testing

Although optical technology continues to advance, it is still possible for a manufacturer or a user accidentally to introduce defects into an objective. If one is unwise enough to disassemble a high-power objective, and simply to rotate the lenses at random about their common axis, a surprising galaxy of aberrations may appear when the objective is put together. This is because the very small, high-curvature elements are not perfectly symmetrical, and the manufacturer has assembled them by trial and error so as to cancel out the defects (the correct orientation should be indicated by marks on the lens mounts). It therefore behoves the customer buying an expensive microscope to make his own tests, at least on the highest-power objective. Other defects may arise from ageing of the cement between lenses.

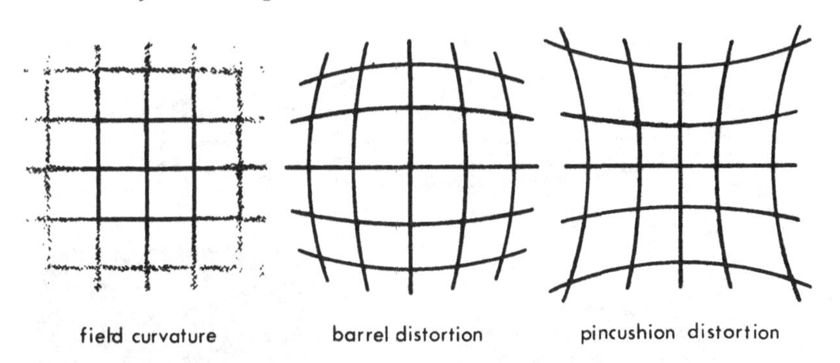

Fig. 47. Field distortions.

A very old method of testing the resolution of an objective is to examine with it a suitable diatom. The skeletons of these tiny organisms, of which there are many kinds, contain extremely regular periodicities, some of which are close to the theoretical limit of the microscope. One can buy slides containing suitable diatoms for a very modest sum. They are, in addition, perfect natural objects for illustrating the Abbe theory (Chapter 4), for the periodicities give well-defined spectra in the back focal plane when the illumination is stopped down (Fig. 48). If one can clearly see a periodicity close to the theoretical limit, there is not much wrong with the objective.

A more precise method of determining which aberration, if any, is present is to examine the Airy disc, seen when the specimen is a slide carrying a vacuum-deposited film of aluminium in which there is a minute pinhole (Chapter 3). If there are no aberrations, the pattern should change in more or less the same way as one moves either above or below focus. When spherical aberration is present this is not so, and the outer rings become more pronounced when 'in focus.' When coma or astigmatism is present, other characteristic changes are seen (Fig. 49). The images above and below focus can be interchanged, depending on whether the objective is under- or over-corrected.

Microscope component markings

In view of the variety of optical components made for different purposes, it is important to understand the maker's distinguishing marks. These vary somewhat from one make to another, and the manufacturer's handbook should be consulted. However, the kinds of thing that one should look for when setting up are as follows:

Condensers vary in numerical aperture (N.A.), and the N.A. is often

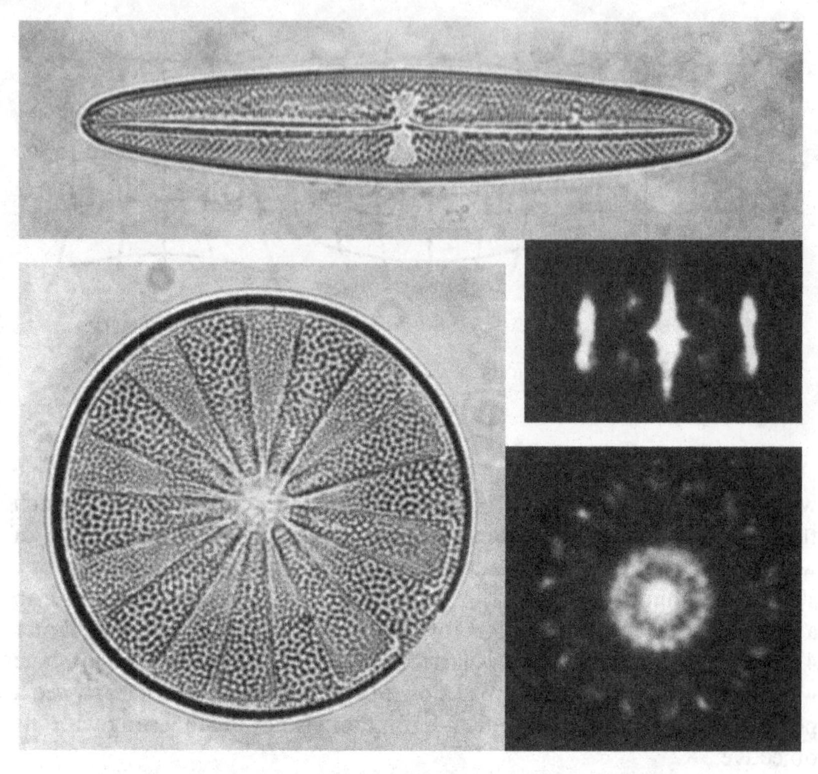

Fig. 48. Diatoms and their spectra in the back focal plane.

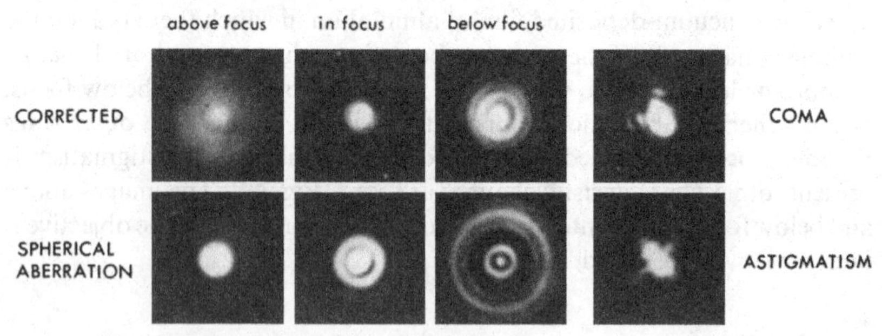

Fig. 49. Changes in the Airy disc pattern.

stamped on the component. It is important to use one whose N.A. is at least as great as that of the highest-power objective to be used, otherwise it will be impossible to obtain an adequate cone of illumination. There are special condensers with a long working distance for use with thick specimen chambers. To achieve the maximum N.A., some condensers need to have a film of immersion oil between the top of the condenser and the specimen slide. Some microscopes have an auxiliary condenser lens for use with certain objectives.

Objectives are usually marked with the magnifying power, but in older microscopes the focal length was given; the power is roughly equal to the distance from objective to eyepiece divided by the focal length. A phase objective may also carry a number identifying the phase ring that must be selected on the condenser mount. The numerical aperture may also appear, together with the tube length in millimetres. This is related to the distance from objective to eyepiece, and may vary between makes; an objective is designed to work best only at a fixed tube length. Finally, high-power 'dry' objectives may carry a number such as '0.17'. This gives the optimum coverslip thickness in millimetres for correction of spherical aberration. The value quoted corresponds to No. $1\frac{1}{2}$ in the British scale of thickness.

Eyepieces must be compatible with the objectives used, and should carry appropriate markings. As mentioned earlier, some are designed to compensate for the aberrations introduced by certain objectives, so it is important to consult the manufacturers' data. Spectacle wearers should if possible choose eyepieces designed specially for a longer working distance; although people suffering only from long or short sight may be able to compensate by removing their glasses and adjusting the focus, those suffering from astigmatism cannot do so. The problems of focusing for photography have already been mentioned in Chapter 1.

Questions

1. In the electron microscope, where a hot filament produces a beam of electrons that is accelerated and then focused by magnetic lenses, what might give an effect analogous to chromatic aberration?
2. A high-power 'dry' objective must have the right coverslip thickness to correct for spherical aberration, but an immersion objective is marked with no recommended thickness. Why is this?
3. What could cause an objective to give an astigmatic image for points *on its axis*?
4. How would you describe, in scientific language, the barrel and pincushion distortions shown in Fig. 47?

10 Photomicrography

Some of the most expensive microscopes incorporate automatic devices for photography, but many users who cannot afford such luxury also need to be able to record what they see. This chapter discusses the problems associated with taking photomicrographs with inexpensive equipment.

Although a normal camera can be used without modification in place of the eye, the relatively restricted field of view of a microscope means that only a part of the camera field may then be filled. The limited aperture of the microscope also means that there is no advantage in using a high-aperture camera lens. One can, however, replace the eyepiece by a *projection eyepiece* which, instead of forming a virtual image (Fig. 2, p. 5) produces an enlarged real image on the film without any intermediate lens. The same effect can be achieved by slightly raising a normal eyepiece to increase its separation from the objective, or by refocusing the objective to change the position of the intermediate real image.

Focusing the image

Focusing on the film is not as straightforward as it might appear. The most direct method is to insert a ground-glass screen in place of the film, and to examine the projected image with a lens. However, the level of illumination is often too low to make this feasible, and the screen introduces a granularity that obscures fine detail. Similar problems can arise if a conventional reflex camera is used.

An alternative procedure is to adjust the microscope by eye with a normal eyepiece, and then to change over to a preset projection attachment. The problem here is that not every person uses the same setting for observing an image: some people relax their eyes and adjust the microscope so as to view the virtual image at infinity, while others unconsciously set the image at some finite distance from the eye. The problem can, however, be overcome if the normal eyepiece incorporates crosshairs or a graticule; the user first adjusts a movable lens to focus on the graticule, and then adjusts the objective focus to bring a sharp image into the same plane. A sensitive test for achievement of this condition is to look for the absence of *parallax*: when the eye is moved from side to side there should be no relative movement of graticule and image.

Choice of film and exposure time

For reasons discussed in Chapter 3, there is no point in using a very fine-grained film for photomicrography; the limit to resolution is generally set by the numerical aperture of the objective. It is more important to choose both film and developer to give a fairly high-contrast result, since microscopic images are often of rather low contrast. A 35 mm camera using medium-speed film is suitable for most purposes, and with black-and-white film the user can quickly develop a series of pictures at minimal cost.

It is preferable to project the image onto the film at somewhat less than the maximum useful magnification (Chapter 3) and to make enlarged prints later, because a shorter initial exposure is thereby required. As an example, a $\times 100$ objective could be used with a $\times 4$ projection eyepiece, followed by photographic enlargement $\times 2$ to the useful limit of $\times 800$ overall; a $\times 8$ eyepiece would demand four times the exposure, because the projected image area varies as the square of the magnification. An image at an overall magnification of $\times 400$ can show no meaningful detail smaller than about 0.1 mm, and this is much greater than the grain size of a medium-speed film.

For routine recording of prints not intended for publication there is a lot to be said for an instant-picture camera of the Polaroid type. However, the fastest films of this kind have a coarse grain structure that may well impair the resolution, and their contrast range is limited. This makes the exposure time more critical than with conventional film. Some Polaroid film packs produce a negative as well as an instant print, and if the exposure is chosen to give good contrast this will be suitable for making enlargements. The exposure may in this case need to be somewhat longer than that required to produce a good instant print.

A common attachment for photomicrography incorporates a beam splitter which sends some of the light to an eyepiece, some of it through a projection lens and some to a photoelectric device for estimating the required exposure. This last may also control automatically the opening of the camera shutter. However, it is often still necessary to try a range of exposures to get the best result. For dark-field images a photometer is unreliable, because it generally measures only the average level of illumination over a given area.

When carrying out trial exposures it is normal to make each one of a series twice as long as its predecessor. Exposure times will be longer when filters are used to produce monochromatic light, and it is worth testing whether the advantages of this (reduction of aberrations and higher theoretical resolving power) are significant in practice.

Factors affecting image quality

The aberration affecting photomicrography most seriously is field curvature (Chapter 9); although a visual observer can adjust the focus to look at different regions of a curved field, photography at high magnifications may demand the use of a special flat-field objective, or appropriate compensation by the eyepiece. Manufacturers' data should be consulted for information on this, and care should be taken to choose a projection eyepiece compatible with the objective.

A more mundane but equally serious problem is that posed by dust particles. Out-of-focus images of such objects, with associated diffraction halos, are easily ignored when viewing by eye but they appear with distressing clarity on a photograph. The problem is accentuated when the numerical aperture is restricted, and a 'shadow image' of a particle situated far away from the object plane may well be obtrusive. It may be necessary to dismantle both the eyepiece and the condenser, and to clean each lens surface individually; care must of course be taken to reassemble all the components the right way round, since many of the lenses have different curvatures on their two faces. It is inadvisable and generally unnecessary to disassemble an objective, though the outer surface of the front lens should be cleaned.

With the highest-power objectives the required exposure time may be inconveniently long: vibration of the apparatus or movement in a living preparation can then blur the image. It may be necessary to change to a high-output light source such as a mercury-vapour discharge lamp, or to use a flash unit. It is in any case desirable to ensure that the camera is rigidly clamped to the microscope body, and to use a release cable for the shutter.

At very low magnifications it may be difficult to achieve uniform illumination of the field, which is again more noticeable in a photograph than in the visual image. An auxiliary condenser lens is supplied with some microscopes to counteract this problem.

Finally, the magnification of a photographic system may need to be measured, and this is best done by taking a photomicrograph of a calibrated stage micrometer or graticule. For accurate work this should be done in several regions of the object field, since even a perfectly-focused image may vary in magnification across the field (Chapter 9). It is also useful in routine recording to have a graticule in the projection eyepiece, so that calibration marks appear in any enlarged prints that are made.

11 Summary, class experiments and further reading

We have seen that all light microscopes have a similar theoretical limit to their ability to see fine detail and distinguish closely-spaced objects, though for the *detection* of isolated objects there is no such limit. For particle counting and the study of long, thin objects such as flagella there is therefore a good case for using the old technique of dark-field illumination, with a high-power light source and a condenser of high numerical aperture; a very high degree of contrast can then be obtained. When it comes to studying more complex structures that are transparent a more modern system must be used, and the choice may turn on how much money is available for purchasing an instrument.

Phase-contrast microscopy is comparatively cheap, and for thin objects it gives contrast as good as that obtainable by any other method. However, it suffers from the 'halo effect' inherent in its design, and for thicker objects the contrast is reduced or even reversed, so that a confusing image is produced. The more expensive interference contrast devices give a clearer impression of the overall mass distribution, and have the further advantage of introducing colour contrast between regions of different thickness. The images also show a 'shadowing' effect, though this is strictly an optical artifact.

The true interference or interferometer microscope is the only one capable of making quantitative measurements of mass. It can also be used for measuring the thickness of a thin film. As the most expensive type available, its use in other applications is something of a luxury.

Polarizing attachments enable one to study macromolecular orientation in a wide variety of objects, and the method has not been superseded by electron microscopy; oriented regions can be detected even when they are embedded in a large excess of non-birefringent material. This is due partly to the dark-field technique associated with the method. The same advantage applies in fluorescence microscopy, where it is now possible to combine great sensitivity of detection with extreme selectivity in the substance labelled.

In purchasing a microscope, therefore, one needs to bear in mind the many applications that are possible. There is a lot to be said for buying a standard 'body' to which all attachments can be fitted.

Class experiments

A brief description is given here, for the benefit of do-it-yourself enthusiasts and anyone setting up a practical class, of experiments that can usefully be done by individuals or small groups. They are designed to bring home the principles discussed in the book, rather than to illustrate the full range of applications of light microscopy.

Basic optics

There is no need to invest in an expensive optical bench. An adequate substitute can be made from a 1 m length of 25×25 mm box-section steel, of the kind used to construct supporting frameworks for benches and laboratory apparatus. Metal or plastic slides are then constructed to hold lenses and other items on a common axis parallel to the bench. The source of illumination can be a tungsten lamp (complete with holder) removed from a microscope, provided that the bulb has a compact filament. A U-shaped metal hood is desirable to shield the lamp. With the aid of inexpensive biconvex lenses, one of focal length about 80 mm and a second of greater diameter and focal length about 40 mm, the following experiments can be done:

Focal length determination. The 80 mm lens is mounted on a slider and moved towards the source until it gives a parallel beam of light, as seen on a white card moved up and down the bench. The distance from source to lens is then measured with a metre rule, and taken as the focal length. This is checked by using the lens to form an image on the card of a distant source of light.

Real image formation. The same lens is used to form an image of the lamp filament on a card held in a second slider. Images are formed for various sets of values of image and object distances u and v (Fig. 1, p. 3), and a rough plot made of u against v. The focal length is calculated from the standard formula (Fig. 1) for a given pair of values, and compared with that determined earlier. Particular note is made of what happens to the image and its magnification when (i) $u < f$ or $v < f$; (ii) $f < u < 2f$; (iii) $u = 2f$; and (iv) $u > 2f$.

Virtual image formation. A piece of thin, preferably semi-transparent graph paper is mounted on a slider and placed 250 mm from the observer, who looks down the bench towards the source. The lamp is moved up to illuminate the paper from behind, and a lens is placed (on its slider) close to the eye. The lens is then moved slowly towards the graph paper until (with *relaxed* eyes) the observer sees a magnified image of the graph paper. By moving the head from side to side one can then estimate the magnification by counting squares on the paper. This is compared with the value calculated from the focal length, using the formula in Fig. 1 (d) with $a=250$ mm.

Microscope illumination. A microscope slide containing a stained section is mounted on a slider about 50 mm from the lamp, so that the section lies on the common axis of lamp and lenses. A card screen is placed at the far end of the bench, and the 80 mm lens moved to form a real image of the section on it. There will be a relatively poor level of illumination. Next, the microscope slide and the lens are removed and the larger 40 mm lens is mounted to form a parallel beam of light as described earlier. The microscope slide is then set up close to the 40 mm lens and on the opposite side to the lamp, and the 80 mm lens set up to form an image of the section as before. Its illumination should be much improved. The effect of stopping down the 'condenser' can be tested with a card containing circular holes. One can also observe on a screen the image of the lamp filament at the back focal plane of the 80 mm lens (see Fig. 2, p. 5).

The compound microscope. In the arrangement just described the 40 mm lens is removed without disturbing the other components; a faint image of the section should still appear on the screen. The screen is now removed and replaced by the 40 mm lens, through which the observer looks towards the source. A badly-focused image of the section should be seen; if not, the vertical or the lateral (sideways) position of the 40 mm (eyepiece) lens may need adjustment. The image can then be focused by moving either of the lenses. If the image is still badly distorted, this may be because the 80 mm (objective) lens is tilted with respect to the optical axis so as to introduce aberrations. The overall magnification can be calculated from measurements made on the individual lenses.

Diffraction phenomena. These can be studied by making suitable masks; in most cases, reduced reversals can be photographed on 35 mm film from drawings made with black ink on white paper. Suggested masks are (i) a single slit of width 0.1 mm; (ii) a slit of width 0.5 mm; (iii) a circular aperture (for this a pinhole in a card is adequate); (iv), (v) pairs of 0.05 mm slits separated by 0.2 mm and 0.7 mm, used in conjunction with the first single slit to reproduce Young's experiment (Fig. 12, p. 19); (vi) a diffraction grating containing a large number of narrow, parallel slits at 20 lines per mm; (vii) a two-dimensional grating made from nylon 'bolting cloth' of the kind used for filtration, having about 15 threads per mm. All diffraction patterns should be observed both in white light and with a green filter. They can be studied either through a lens or with the naked eye, holding the mask close to the eye and some distance from the lamp.

Phase contrast and dark-field microscopy

Bacillus megatherium is a suitable test object for comparing the various methods of visualizing small, transparent objects. The bacilli mentioned are non-pathogenic and conveniently larger than most species. They should be grown up overnight to the vegetative form on an agar slope inoculated with spores (obtainable in the UK from the National Collection of Industrial Bacteria, Torry Research Station, Aberdeen); the surface of the slope is then washed with a small amount of distilled water, and drops of liquid spread on a number of clean microscope slides with a glass rod. When dry, the slides are dipped in methanol to fix the bacteria and remove soluble salts. They can then be mounted in media of various refractive indices, both above and below that of the bacteria themselves.

The phase-contrast intensity curve shown in Fig. 25, p. 35 can be illustrated by making a slide with a small drop of human blood (obtained by pricking an alcohol-cleaned finger with a sterile lance), diluted with physiological saline. The erythrocytes are observed under low-power phase contrast, and a drop of 0.5% saponin added to one edge of the coverslip. As the saponin diffuses into the slide, the erythrocytes swell up and then haemolyse; the contrast they show against the background changes in a way that is fully consistent with Fig. 25.

Fluorescence microscopy

If it is not practicable to demonstrate antibody fluorescence, the acridine orange effects described in Chapter 6 can be shown with a culture of *Chlamydomonas* or *Tetrahymena*.

Polarizing microscopy

Urea crystals can be used to illustrate the principles of birefringence measurement (Chapter 7); they have recognizable long and short axes, and there are different refractive indices for light polarized along these axes. One needs to choose a small crystal (or a fragment obtained by crushing large ones) that shows a strong, uniform colour when viewed between crossed polarizer and analyzer. A fibre of deoxyribonucleic acid (DNA) is also suitable; salt-free DNA is wetted with distilled water and the gel stretched while drying, to give a fibre of diameter about 0.1 mm.

Having determined the sign of birefringence of DNA one can determine the orientation of DNA molecules in a suitable biological specimen, such as a preparation of sperm from a squid. These are examined with the first-order red plate. Other more easily prepared small objects, which show an interesting pattern of different colours with a first-order red plate, are the starch grains obtained by rubbing a cut potato on a slide. They are best mounted in glycerol; the starch molecules are positively birefringent chains, and knowing this one can work out how they are arranged within the grain. A less complex problem is provided by a human hair, which again contains positively-birefringent macromolecules. A fair hair is best, again mounted in glycerol to reduce refraction at the surface.

Interference microscopy

Bull sperm heads are suitable objects for the determination of mass. The area in projection, A, is estimated by assuming an elliptical form and measuring major and minor diameters a and b; $A = \pi ab/4$. The mass per unit area is determined as described in Chapter 8.

Pollen grains are suitable objects for demonstrating mass contours, provided that they are not too rough-surfaced. A slide of an oil–water emulsion also shows excellent contours in each droplet. For thickness determination an electron microscope section can be used, or a very thin plastic sheet; however, the latter should not show appreciable birefringence. Domestic 'cling film' (r.i. 1.53) is suitable if stretching is avoided.

For qualitative observations by interference contrast, pollen grains are again suitable. Human epithelial cells, obtained by scraping the inside of the cheek with a clean fingernail and mounting in saliva, are of greater interest because a number of subcellular structures can be seen.

Further reading

The short list below includes some basic texts which do not 'date', and some that describe recent developments.

Bradbury, S. (1968) *The microscope past and present*. Pergamon, Oxford.

Glauert, A.M. (ed.) (1974) *Practical methods in electron microscopy, Vol. 2*. North-Holland, Amsterdam.

Hartshorne, N.H. and Stuart, A. (1970) *Crystals and the polarising microscope*. Arnold, London

Jenkins, F.A. and White, H.E. (1976) *Fundamentals of optics*. McGraw-Hill, New York.

Lipson, S.G. and Lipson, H. (1969) *Optical physics*. Cambridge University Press.

Martin, L.C. (1966) *The theory of the microscope*. Blackie, London.

Meek, L.C. and Elder, H.Y. (1977) *Analytical and quantitative methods in microscopy*. Cambridge University Press.

Slayter, E.M. (1970) *Optical methods in biology*. Wiley-Interscience, New York.

Useful booklets which include some basic theory are also published by certain manufacturers, notably Zeiss (West Germany) and Leitz.

Appendix: Mathematical treatments

Energy carried by a wave

The energy carried by a single wave whose electric field is represented by $E = a \sin \theta$ is proportional to the square of the amplitude a. To understand why this is so we can imagine the wave passing a detector comprising a pair of metal plates like those in Fig. 3 (p. 10); instead of a fixed field being produced by a battery as in Fig. 3, the travelling wave now induces a sinusoidally-varying voltage between the two plates. The voltage is directly proportional to the field, which is defined in terms of voltage per unit distance.

If in place of the battery we now connect a length of electrical resistance wire, the voltage induced by the field will cause a current to flow through the resistance. This will cause it to heat up, so that energy is extracted from the wave; in principle all the energy carried by the wave could be used up in this way. Something analogous to this happens at the molecular level in a light-absorbing substance.

The rate of extraction of energy from the wave will depend on the product of current and voltage (the wattage), and since current is proportional to voltage (Ohm's law) this means that the wattage is proportional to the square of the applied field. The average rate over a whole cycle of the sinusoidal fluctuation in field is then proportional to

$$\int_0^{2\pi} E^2 \, d\theta = a^2 \int_0^{2\pi} \sin^2\theta \, d\theta = \pi a^2$$

The intensity I of a light wave is defined as the square of the amplitude a, so this means that I is proportional to the energy carried by the wave.

Coherent and incoherent sources

When any two sinusoidal waves of the same frequency interfere, their resultant amplitude is given by the vector diagram of Fig. 9 (p. 15). The mathematical expression of this is

$$a_R^2 = a_1^2 + a_2^2 + 2a_1a_2 \cos \phi$$

where a_1, a_2 and a_R are the amplitudes of the waves concerned. The energy carried by the resultant wave is therefore proportional to a_R^2, as explained in the previous section, provided that the phase difference ϕ between the waves does not change during the cycle. Waves of this kind are *coherent*, and clearly one must obtain the vector sum of the fields before squaring the amplitude. Depending on the value of ϕ, this may be either greater or less than the intensity sum $(a_1^2 + a_2^2)$. We shall see later that this does not imply any overall gain or loss of energy.

When the two sources of light are *incoherent*, their waves have a phase difference that fluctuates randomly with time; although each photon from a given source is a coherent 'packet', the next one will have a phase unrelated to that of its predecessor. Waves of equal amplitude from two different sources may therefore, at any instant, either reinforce or cancel each other. The resultant intensity is then obtained by averaging over all possible values of ϕ, so that the expression above becomes

$$I_{av} = \frac{1}{2\pi} \int_0^{2\pi} a_R^2 \, d\phi$$

$$= a_1^2 + a_2^2 + \frac{a_1 a_2}{\pi} \int_0^{2\pi} \cos \phi \, d\phi$$

The integral of $\cos \phi$ from 0 to 2π is equal to zero, so I_{av} is equal simply to the sum of the incident-wave intensities.

Young's experiment

The path AP travelled by one of the waves in Fig. 12 (p. 19) is given by

$$AP^2 = a^2 + (x - \tfrac{1}{2} y)^2$$

so that

$$AP = a\{1 + (x - \tfrac{1}{2}y)^2/a^2\}^{\frac{1}{2}}$$

Since x and y are in practice much smaller than a we can expand by the binomial theorem and write

$$AP \approx a + (x - \tfrac{1}{2}y)^2/2a$$

Similarly for the other wave

$$BP \approx a + (x + \tfrac{1}{2}y)^2/2a$$

The path difference is therefore

$$BP - AP \approx xy/a$$

When this is equal to a whole number of wavelengths there will be reinforcement, and a bright fringe will result; the condition for this is

$$xy/a = N\lambda$$

where N is an integer, or

$$x = N\lambda a/y$$

If A and B are apertures of equal width, and they are equally illuminated by the source, the intensity will fall to zero between each fringe. On the other hand, the intensity at the centre of a bright fringe will be twice what it would be if the two effective sources A and B were incoherent. This does not imply a net gain in energy but simply a redistribution: integration shows that the total energy reaching the screen is the same as if A and B were incoherent sources.

The Airy disc

Diffraction from a circular aperture presents a complex mathematical problem, but the simpler case of an infinitely-long slit of constant width will illustrate the method of arriving at a solution. Let the diagram on the left-hand side of Fig. 13 (p. 20) be taken to illustrate such a slit, rather than a circular aperture. The lens brings to a focus all the light diffracted in a given direction, and the resultant amplitude at a point in the focal plane is obtained by summing the contributions from all regions of the aperture.

We can avoid a full analysis by considering pairs of narrow elements (Fig. 50) separated by *half* the aperture width d; the aperture can then be considered as made up of a series of such pairs of imaginary slits, each shifted slightly with respect to its predecessor. The path difference between waves from any given pair of elements is then $\tfrac{1}{2}d \sin\theta$. It is clear that when $\sin\theta = \lambda/d$ the path difference will be half a wavelength, and the two waves when combined by a lens will cancel out; since this applies to every other pair of waves as well, the integrated amplitude from the whole aperture must be zero. This corresponds to the first minimum, at angle α, in the pattern of Fig. 13. The only effect on this result of substituting a circular aperture for the slit is to make the condition $\sin\theta = 1.22\lambda/d$. The derivation of this involves Bessel functions and is beyond the scope of the present book.

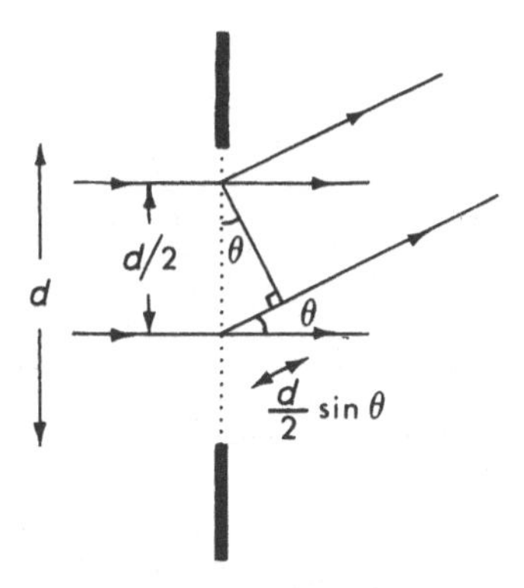

Fig. 50. Diffraction from elements of an aperture.

The resolution limit

From the analysis given above it is apparent that, so far as the position of the first minimum is concerned, a slit of width d is equivalent to a circular aperture of width $1.22d$. From Fig. 50 we can see that when the path difference between waves from pairs of elements is $\frac{1}{2}\lambda$ (as it is for the first minimum), the path difference between waves from the two edges of the slit is λ. For an equivalent circular aperture, therefore, the path difference between 'edge waves' is 1.22λ.

Applying this to the situation in Fig. 15 (p. 22) we see that the condition for the first minimum in the dotted curve to overlap the central maximum of the solid curve is given by $\Delta = 1.22\lambda$, where Δ is the optical path difference between waves travelling to the focal image of the first minimum via the two extremes of the lens diameter. Since in Fig. 15 the point of overlap lies on the axis of the lens, the path difference can in this case be regarded as arising solely to the left of the lens: it is due to the off-axis position of the corresponding object point.

Fig. 51 shows that while waves from the axial point A have equal paths to the lens extremes C and D, the two 'edge waves' from point B have paths BC and BD which are different. BC is shorter than AC by a distance of approximately $y_R \sin \theta_0$, while BD is longer than AD by the same amount. The geometrical path difference between BD and BC is therefore $2y_R \sin\theta_0$, and the *optical* path difference is $2ny_R \sin\theta_0$ where n is the refractive index of the surrounding medium (Fig. 15).

The resolution limit is therefore obtained by putting this o.p.d. equal to $1.22\lambda_0$, where λ_0 is the wavelength in free space (not the immersion medium). We then have

$$y_R = 0.61\lambda_0/n \sin\theta_0$$

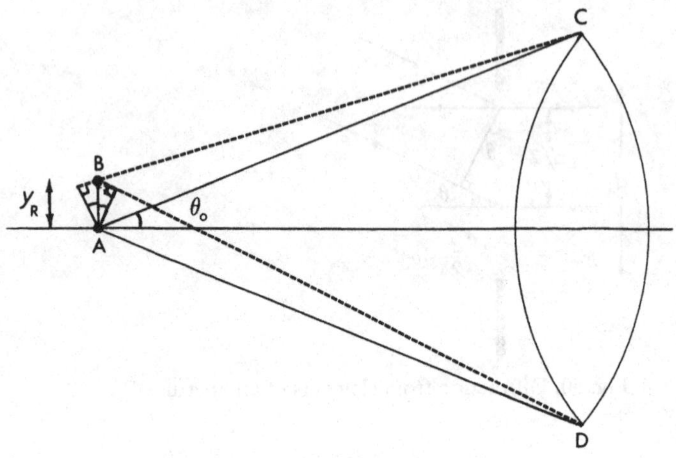

Fig. 51. Object points at the resolution limit.

Fourier's theorem

Any single-valued function of a variable x can be expressed as a series involving the sines and cosines of integral multiples of x, e.g.

$$\begin{aligned} f(x) = {} & b_0 + b_1 \cos x + b_2 \cos 2x + \dots \\ & + a_1 \sin x + a_2 \sin 2x + \dots \end{aligned}$$

or
$$f(x) = \sum_{m=0}^{m=\infty} (b_m \cos mx + a_m \sin mx)$$

A function that has discontinuities, or is non-periodic, will require a large number of terms to represent it adequately. The term b_0 represents a constant background level as in Fig. 16 (p. 25). Certain terms or groups of terms may have zero amplitude; for instance, a function with $f(-x) = f(x)$ has no sine terms, while some functions lack even- or odd-order terms. The amplitudes of the component terms can be derived from $f(x)$ by integration procedures described in books such as that by Martin (Chapter 11).

Answers to questions

Chapter 1

1. By imaging a *distant* source (such as the sun, a room light or a lighted window) onto the card. The focal length is then close to the separation of lens and card.
2. Magnification = $250/f$, where f is the focal length in millimetres. 250 mm is taken as the minimum viewing distance for the naked eye.
3. There is a virtual image at infinity, and a real image on the retina of the eye.
4. No, but the image will be unevenly illuminated.
5. No images are formed, but scattered light may contribute to the background level and reduce the contrast.

Chapter 2

1. Some of the wave's energy is absorbed, changing permanently its amplitude and therefore the brightness. The wavelength also changes, but only while the wave is within the specimen.
2. Transmission of energy from source to observer.
3. They are directional: the field exists only in a plane at right angles to the direction of propagation.
4. It decreases in the same proportion.
5. No change, because this rate is determined by the source of the wave.
6. Again no change, since colour is determined by frequency which does not alter; the colour detectors in the eye are sensitive to certain ranges of frequency, and do not respond to changes of phase.
7. 90° or 270°.

Chapter 3

1. The pinhole, which being smaller gives a more extended diffraction pattern.
2. Because the diameters of the rings depend on the wavelength, so that each component colour gives a different pattern.
3. It gets smaller, because the wavelength of blue light (about 470 nm) is smaller than that of green (about 520 nm).
4. No, because the resolving limit is determined by the objective. There is no

point in using film whose grain size is much smaller than that of the detail in the image; it will also require a longer exposure.

Chapter 4

1. The grating periodicity will appear to be halved, because the second-order spectra have the same angle of diffraction as the first-order spectra of a grating with twice as many repeats per unit length.
2. Again there will be an apparent 'halving' of the repeat. The effect is as if, in Fig. 16, we squared the amplitude of the $m=1$ component to obtain the intensity; both positive and negative halves of the amplitude cycle would then give bright fringes.
3. No, because the numerical aperture would be too low to accept any diffracted spectra; the first-order spectra are only just collected by the $\times 100$ objective.
4. Because it involves the same principle of reconstructing an image from diffracted spectra; only the methods are different.

Chapter 5

1. Because this causes a loss of resolution and distortion of the image, due to removal of higher-order spectra.
2. When its thickness is uniform and it gives a phase shift corresponding to the 'cross-over' in Fig. 25, diagram 5.
3. By changing the refractive index of the medium to alter the relative phase shift introduced by the object.
4. By immersing the object in a series of oils of different refractive indices (r.i.); when the r.i. of the oil exceeds that of the object the contrast will reverse, and when there is a perfect match it will be invisible.
5. Each point (e.g. along the length of a thin flagellum) interrupts the incident wavefronts, and generates *reflected* Huygens' wavelets centred on that point. The disturbance propagates out in a range of directions *behind* the object, so that with oblique illumination some of this reaches the observer.

Chapter 6

1. Because its spectrum is not continuous, but concentrated into *lines* at certain wavelengths. None of the stronger lines lies in the excitation region for FITC, so one is left with the weaker background spectrum.
2. Very little, since the path lengths in the various layers remain unchanged.
3. Because at oblique incidence the optical path differences between interfering beams will change. The geometry predicts a shift to a shorter wavelength. For this reason, interference filters used in strongly-convergent light will show an increase in the width of the pass-band.
4. The method relies on the purity of the antigen used to raise the antibody. If there is minor contamination by a substance present in the cell in large amounts, the fluorescence from the antibody to this component may dominate the picture obtained. Such problems did arise in early research with the technique.

Chapter 7

1. Yes. The extra sheet passes a component of the light coming from the polarizer, polarized along the new vibration axis. This in turn gives a smaller component polarized along the vibration axis of the analyzer. The explanation is analogous to that for a sailing boat moving against the wind.
2. It will pass through a maximum when the refractive index of the medium is equal to that of the solid phase.
3. The short axis.
4. 5.46×10^{-3}.
5. Because a given o.p.d. will contain a larger number of short wavelengths than long ones. This is nothing to do with dispersion, which has only minor effects (such as the failure to produce a completely black appearance when compensation is achieved with white light).

Chapter 8

1. Because this is essential to the working of the beam splitter and combiner, which bend one beam but not the other. The use of a compensator also relies on the two components differing in polarization.
2. As Δ increases from $\frac{1}{2}\lambda$ to λ the brightness of the background returns to zero, while the 'shadowing' effects are as if the light came from the opposite side of the field.
3. It will become fainter, and the 'shadowing' effects will vary differently with Δ.
4. The final answer is unaffected; although the o.p.d. due to the specimen will decrease, so will the specific refractive increment. For highest accuracy it is desirable to have as large a difference of refractive index as possible, and water ($n = 1.33$) is normally used in biological applications.

Chapter 9

1. A spread in electron velocity, caused by *a*) the range of thermal energies among electrons emitted by the filament; *b*) variations in accelerating voltage and lens currents; and *c*) energy variations due to the slowing-down of electrons by passage through the specimen. Only the last is important in modern microscopes.
2. Because with immersion oil there is hardly any refraction at the upper surface of the coverslip of the kind shown in Fig. 45; the coverslip therefore introduces no aberration.
3. Asymmetry of optical surfaces, or faulty assembly of a lens so that it is tilted in its mount.
4. The image is everywhere in focus, but the magnification varies with distance from the centre of the field.

Index